The Geopolitics of Energy

Other Titles in This Series

8

Westview Special Studies in Natural Resources and Energy Management

The Geopolitics of Energy
Melvin A. Conant and Fern Racine Gold

How can the industrialized countries reduce their vulnerability to supply disruptions caused by continued dependence on foreign sources of oil? How can access to Middle East oil be made more secure? These are the core questions that arise from a new worldwide energy situation in which the industrialized countries have remained dependent on oil and oil imports for their economic, political, and military well-being, while control of these resources has passed to an increasingly small number of less-developed countries whose interests do not automatically or necessarily coincide with those of the consuming, industrialized countries.

With a focus on these questions, *The Geopolitics of Energy* analyzes the present worldwide energy situation and its likely evolution over the remainder of the century. The authors consider likely developments in coal, gas, and nuclear energy; the outlook for oil, which will remain the dominant energy source at least through the 1990s; and the implications of this energy outlook for U.S. foreign policy, intra-Western alliance relations, and North-South and East-West relations. Identifying the issues that will concern governments as long as the need for oil is pervasive—until alternative energy sources begin contributing significantly to world energy supply—the authors conclude with policy recommendations for the United States based on their analysis of the energy situation and its consequences. This book is based on a report prepared for the U.S. Department of Defense.

Melvin A. Conant, an international energy consultant, was previously assistant administrator for international energy affairs in the U.S. Federal Energy Administration and senior government relations counselor for the Exxon Corporation.

Fern Racine Gold is an international energy consultant on oil and nuclear supply and a doctoral candidate at Columbia University. She was previously a political analyst for the Exxon Corporation.

"Oil will remain a source of political and economic power for the exporters until the end of this century, at least. . . . World oil consumption has begun to rise, and will increase as the years pass. Oil will retain its role as a source of energy."

—Prince Fahd ibn 'Abd al-'Aziz,
Crown Prince and First Deputy
Prime Minister, Saudi Arabia,
July 2, 1976

The setting of oil prices and production levels remains a "non-negotiable sovereign right" of OPEC states. If their objectives are not achieved, "prices will not be the only problem; availability of oil supplies will also come into question."

—Dr. Abdul Tahy Tahar, Chairman,
Pertamina, Saudi Arabia, May 1977

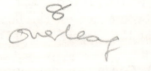

The Geopolitics
of Energy

Melvin A. Conant
Fern Racine Gold
Foreword by Robert Ellsworth

Westview Press • Boulder, Colorado

Westview Special Studies in
Natural Resources and Energy Management

Published in 1978 in the United States of America by

Westview Press, Inc.
5500 Central Avenue
Boulder, Colorado 80301
Frederick A. Praeger, Publisher and Editorial Director

Library of Congress Cataloging in Publication Data
Conant, Melvin.
    The geopolitics of energy.
    (Westview special studies in natural resources and energy management)
    Includes bibliographical references.
    1. Power resources. 2. Energy policy—United States. 3. World politics—1975-1985. I. Gold, Fern R., joint author. II. Title.
HD9 502.A2C648 1977b   333.7   77-20668
ISBN 0-89158-404-8

Printed and bound in the United States of America

Contents

Figures and Tables

Abbreviations

CIEC—Conference on International Economic Cooperation
Comecon—Council for Mutual Economic Assistance
EC—European Communities
FEA—Federal Energy Administration
IAEA—International Atomic Energy Agency
IEA—International Energy Agency
LDC—Less-Developed Country
NPC—National Petroleum Council
OAPEC—Organization of Arab Petroleum Exporting
    Countries
OECD—Organization for Economic Cooperation
    and Development
OPEC—Organization of Petroleum Exporting Countries
VLCC—Very Large Crude Carrier

Foreword

An earlier version of this book by Melvin A. Conant and
Fern Racine Gold, his associate—a report prepared in 1976—
has already helped shape both thought and the bureaucracy
in Washington. Of course, that is never a mean accomplishment; in this case it owed as much to the authors' clear and
authoritative projection of reality as it did to Washington's
readiness to pay attention.

The nation for years has felt severely frustrated by
Washington's inability to deal effectively with international
energy problems. President Carter's initiatives have relieved
some of that frustration, but there has been—and still is—
woefully little understanding of the world geopolitics of
energy. It was in order to get at this lack of understanding
that I asked Mr. Conant in 1976 to prepare a study on the
subject for the Department of Defense, of which I was then
Deputy Secretary.

As soon as the report was completed, Senator Henry M.
Jackson—one of the nation's keenest and most influential
statesmen—published the document under the imprimatur of
the Senate Committee on Interior and Insular Affairs, of
which he was then Chairman. (Senator Jackson is today
Chairman of the Senate's new Committee on Energy.)

Now, with Westview's publication of the following book
(an expanded and up-to-the-minute version of the original
report), the wider world of scholars, journalists, and planners,
as well as high officials, all around the world, will be able to

profit from this important work. In it, the authors thorough-ly illuminate those geographical and political patterns that condition and limit international energy relations: patterns that were exposed briefly and quite dramatically at the time of the oil embargo of 1973.

As for the future, the probability is that those who control geography-specific energy such as oil and gas, coal, and uranium will not design their policies entirely to fill the needs of those who will require large and growing amounts of oil for their survival. Therein lie grave risks, which will continue to engage the full range of nations' interests: risks that cut as close to the bone of national security as any other risks from any other source. This book helps us understand these mat-ters, too.

The authority of *The Geopolitics of Energy* derives not only from the scope and clarity of the writing, which the reader will judge for himself, but also from the experience and perspective of the authors. Melvin A. Conant has taught on the faculty of the U.S. National War College, he has con-ducted studies at the Council on Foreign Relations, he has enjoyed over ten years' experience in the international oil industry, and he then helped establish the first significant international energy office within the U.S. government.

Fern Racine Gold, then with the Exxon Corporation, was given leave to apply her own very considerable skills to the project; subsequently, she joined Mr. Conant in undertaking further studies relating to the politics of international energy.

It is hard to see how the reader who is drawn to the title can fail to be fascinated with the book.

Robert Ellsworth
Washington, D.C.

Preface

In 1973, the oil-producing countries unilaterally assumed
control of oil prices and production levels from the international oil companies. *The Geopolitics of Energy* is an analysis
of the consequences of the new worldwide energy situation in
which the industrialized countries remain dependent on oil
and oil imports for their economic, political, and military
well-being, while control of those resources has passed to an
inreasingly small number of less-developed countries whose
interests do not automatically or necessarily coincide with
those of the consuming, industrialized countries.

Part 1 examines the worldwide energy situation and its
likely evolution over the rest of the century. This section of
the book considers developments in coal, gas, and nuclear
energy as well as the outlook for oil, which will remain the
dominant energy source, at least through the 1990s.

Part 2 considers the implications of the worldwide energy
outlook for U.S. foreign policy, Western-Japanese alliance
relations, and "North-South" and "East-West" relations. In
effect, given the energy outlook described in Part 1, what are
the implications?

Issues are identified that will concern governments as long
as the need for oil is pervasive and before alternative energy
sources begin contributing significantly to world energy supply. The fundamental questions are two. How can the industrialized countries reduce their vulnerability to supply disruptions, vulnerability derived from their continued dependence

on foreign sources of oil? How can access to Middle East oil
be made more secure?

The final section of the book includes policy recommendations for the United States based on the analysis of the
energy situation and its consequences.

Part 1

Background to Policy

Introduction

What aspects of energy supply will engage the interests of states from now until the early decades of the twenty-first century? What issues will preoccupy the great industrial states until solar power, nuclear fusion, or some other energy form relieves them of the challenge of securing access to adequate and continuous supplies of energy from foreign sources?

Access to raw materials generally, and to energy in particular, is certain to be a major preoccupation in international political relations. The issues involved reflect a changing international environment in which the availability of basic commodities is no longer guaranteed by traditional colonial relationships or power defined in military terms. In an earlier era, doubts as to access would have been resolved by the action of great powers. Today, the growing unwillingness to use direct military force gives other factors, such as political control of energy resources, greater significance in the calculation of interests. The contemporary international environment tends toward a wider definition and dispersion of the elements of power.

"Access" will be determined by an interplay of geographic factors and government policies based on a complex mix of political and economic considerations whose ingredients will vary state by state. The terms under which those who control resources make them available to those who depend on them will reflect changes in the international environment

and evoke still further changes of great international consequence. Access will not be determined solely by need and
certainly not by an industrialized state acting unilaterally.

These considerations have implications for international
power relationships. Changes in the distribution of power are
likely not only in terms of North-South relations but in the
relative positions of countries in the developed world as well,
including East-West relations and relations among the countries of the Western alliance and with Japan.

In the case of energy, we share a belief that in the early
decades of the next century another profound revolution—
comparable to the one that followed the use of coal and oil—
is to be anticipated. The use of new sources of energy could
free many states from most geographic restraints. Until then,
the interests of great and smaller powers will be engaged in
the pursuit, and wise utilization of energy resources—resources as vital to their societies' well-being as food and
water. If governments fail to use the transitional period to reduce dependence on imported energy and to develop alternatives to oil, competition for available supply will be intensified.

The Geopolitics of Energy

Geopolitics as an approach to the study of international relations stresses the importance of locational factors in influencing relations among nations. Thus, geopolitics emphasizes
geographic factors as important determinants of government
policy and major determinants of the relative power position
of states. In this report on energy, these locational factors are
emphasized, as they must be, in considering access to raw materials generally.

In turn, the importance of various geographic factors
changes with developments in many areas, including the passage of time, advances in technology, the need for access to
raw materials, and changes in national and international political goals and judgments as to legitimate means of pursuing

international objectives.

Moreover, geographic and locational factors vary in importance with changes in the international system itself; there are new international actors (new nations as well as multinational corporations, international organizations, and regional economic and military organizations); the legitimacy and adequacy of the traditional actors, of nation-states themselves, is a subject for debate. Power becomes more widely dispersed; superpowers often find themselves confounded by lesser states who find room for maneuver within the stalemate created by nuclear weaponry. Interdependence, in terms of mutual dependence as well as interpenetration, is a reality. And it is within this continuously changing international environment that geopolitics and access to raw materials will evolve.

Are there geopolitical factors related to energy and raw material supplies that suggest the outline of new international relationships for the decades ahead? Which areas, by dint of their control over which geographic factors, will be strategically and economically important in the future? What combinations of states are made likely by these geographic factors? If energy is of vital interest to the world community, will there be "energy heartlands," other than the Middle East, of undisputed significance, access to which will be of prime importance?

Because all major primary energy resources depend on a host of additional actions necessary to (1) transform them into usable form and (2) transport them to consuming areas, moreover, factors other than resource location are essential aspects of the geopolitics of energy. The logistical supply lines, the technology, and the processing facilities without which the raw resource is of little value will also have implications for international politics.

Finally, the continuous interaction between factors influencing supply (reserves, processing, new discoveries, growing energy consumption, and energy research and technology) and factors influencing demand (economic growth, resource requirements of a particular economic system, and availability of substitutes), which gradually give different resources and geographic factors different importance over time, is also an important aspect of the geopolitics of energy.

National Interest and National Security

Despite efforts to define the energy needs and policies of the United States, we still have no comprehensive, realistic, and generally acceptable national energy policy and plan. The proposals of President Carter are the latest of a series of recent efforts to grapple with energy dependence. We cannot be sure that the final policy and plan will be any more comprehensive than earlier undertakings. We are not alone, for there are very few other industrial states that can claim to have such.

For the United States, energy objectives have been lost in a great many other concerns and in a quagmire of domestic politics. With time, the government has shown a capacity to learn and to acknowledge the implications of its rising dependence upon imported oil. In the not-so-recent past, energy "initiatives" and "policies" reflected tactical actions to secure near-term objectives, a shortsighted approach to a very major national interest.

As it defines and implements a more adequate energy policy, the United States will become more aware of the extent to which degrees of dependence upon energy in international trade affect the selection of allies, modify alliances and, perhaps, create the need for new ones. The geological "accidents" that concentrated energy resources in preindustrial societies bring the issue of overall "North-South" relations into the search for access, a search in which the United States may actually find itself pitted against traditional allies, or pursuing such divergent causes as to affect the durability of our relationships.

Scope of Government Interest

States dependent on imported energy resources have two

cardinal objectives: first, to pursue policies designed to secure access to the additional foreign supply that is essential to their national requirements; and, second, to reduce the necessity for access to foreign supply.

In assuring itself of access to foreign supply, a government has a variety of options: it may seek bilateral relationships with key producers; it may create a system of preferred sources; it may participate in more general undertakings such as the Common Market–Arab dialogue or trade arrangements such as the Lomé Convention; it may join in technological assistance efforts; or it may participate in still wider international efforts such as the International Energy Agency (IEA), the Conference on International Economic Cooperation (CIEC), and commodity agreements. It may, of course, do all of these.

In reducing the necessity for access to foreign supply, a government can implement policies of conservation, give incentives for the development of energy alternatives, encourage research, and the like. These are essentially internal actions that a government may take to reduce demand and encourage indigenous production. The success of these policies will be determined by timely actions, political will, appropriate economic policies and regulation, and the extent to which nature favors the country with energy resources.

The issue of access to energy resources actually involves three interests, which each energy-deficient state shares with all others:

1. A state's supply of imported energy must be *adequate in volume;* there is a level of imports below which national security is jeopardized.
2. The supply of imported energy must also be *continuous.* Interruptions or occasional shortfalls in supply can have serious economic and political implications for industrialized states. It is, of course, this vulnerability to disruptions of supply that gives resource-rich states a lever to use against states dependent on imported energy.
3. Imported energy must also be available at "reasonable" prices—the most difficult to define of the three aspects

of access. Clearly, price should bear some relationship to the cost of alternative forms of energy—both available and prospective; price should also reflect the fact that present energy sources are diminishing and nonrenewable. *Price should also reflect a "capacity to pay."*

These three factors—adequate volume in continuous supply and at reasonable price—constitute an interrelated triad of energy interests. Failure to obtain any one of the three could have disastrous consequences for the economic well-being, political stability, and national security of the consuming country.

1
Highlights of the Contemporary Geopolitics of Energy

The twentieth century has witnessed the greatest shift in energy sources the world may have experienced since the spread of the use of fire. In the first quarter of this century, coal was indisputably the major source of energy for the industrial world. The energy requirements of great states could be met totally from within their borders or supplemented from nearby resources (in the case of Japan). Coal would have remained by far the key energy source had not the discovery of large volumes of oil in southern Russia, the Middle East, and, later, in the United States quickened interest in the comparative ease of its extraction and transportation, and its conversion to meet a host of requirements.

What circumstances produced the revolution in oil beginning in the mid-fifties? What circumstances bring its availability to the highest order of national interests? These may be quickly highlighted by statistical references to the growth in energy consumption, beginning with 1960 (the year the Organization of Petroleum Exporting Countries [OPEC] was founded): in that year, the world's energy consumption was about 132 quintillion British Thermal Units (QBTU); ten years later, it was some 217 QBTU; five years later, in 1975, it is estimated to have been 225 QBTU. Fourteen years from now—in 1990—it could be 415 QBTU—more than a threefold increase in total world energy consumption in only thirty years.

But of those increases, what was the share of oil (and

natural gas)? In 1960 they represented 48 percent of the
world's energy consumption. Ten years later, it was 63 percent; by 1975, it was some 67 percent; and by 1990, even assuming a very large increase in the role of nuclear power, it
could be about 85 percent. The volumetric implications are
staggering. Eight billion barrels of oil were consumed
throughout the world in 1960, 17 billion barrels in 1975,
and possibly 30 billion barrels by 1990: nearly a fourfold increase in the use of oil and gas in thirty years. Coal, on the
other hand, which had been the primary source, was nearly
47 percent of the world's consumption of energy in 1960,
but it sank close to 30 percent in 1976.

The convenience of oil, therefore, its scant labor requirements, its extraordinary range of uses, and perhaps most important of all, its relative cheapness, plus the enormous expansion in producing capacity and huge reserve discoveries—
all combined to make it and its products the most attractive
and primary source of energy.

The key decision that catapulted oil into what would
eventually be energy's first place came with the pre–World
War I undertaking by the British Admiralty to convert its
battle fleet to oil, a decision quickly followed by every major
power. A whole set of geopolitical factors emerged from this
far-reaching commitment: access to oil imposed new and
greater commitments on foreign and defense policies. For the
British especially, given the size and role of the Royal Navy,
the Middle East, which was still considered the "bridge" to
India and the East, a bridge to be defended against Russian
ambitions, now acquired an additional strategic purpose: access to, and protection of, the oil fields of Persia and the
Gulf.

Before World War II, French, German, and American commercial interests also sought access to that oil. While the strategic importance of oil may have been influential, it was not
necessarily the predominant factor in impelling countries to
move into the area. In the German case, the desire for general
strategic advantage vis-à-vis Britain may have outweighed considerations of oil per se; similarly, France's traditional rivalry

with Britain may have provided the original reason for a French presence in the Middle East, which in any case was not predominantly to meet a French need for oil. For the United States, commercial interests dominated. Perhaps mainly in the case of Japan (which, however, looked to Southeast Asia for supplies) did a recognition of the strategic importance of adequate oil supplies motivate foreign activities.

After World War II, the threat of Soviet expansion into the Middle East and the creation of Israel added new dimensions to U.S. interests. The increasing importance given to oil in world energy trade rapidly expanded the catalog of U.S. concerns. Nevertheless, the United States did not debate the longer-range implications of this greater stress on oil generally, nor its exclusive access to the oil of Saudi Arabia. This is still true, although allies and others find it disingenuous of the United States to continue explaining that there is no "special relationship" regarding the immense oil resources of that kingdom, on whose policies and actions the energy interest of so many depends.

Nevertheless, industrial nations generally have not had energy policies appropriate to the extent of their dependence. It is not surprising, therefore, that by the beginning of the 1960s, when oil consumption began truly to soar and to rank high among any country's strategic interests, political developments among the producer countries shattered the imperial system; control of oil passed to the other side. The change has been too rapid to allow time for careful consideration of the policy alternatives necessary to regain the requisite assurance of supply.

Energy Needs of the Industrial World (1960-1977)

The exponential growth in the imported energy requirements of the leading industrialized states is the basic condition that initiates a discussion of the geopolitics of energy.

Three sets of data display the situations: (1) the increase in energy consumed by these states, (2) the extent to which these needs have been met by oil and through oil imports, and (3) the importance of the Middle East and Africa as the source of supply. From table 1, note that

1. Over the period 1960-1975, energy consumption doubled generally and tripled for Japan; Europe and Japan's dependence upon imported oil remained, for all practical purposes, total.
2. From 1960 to 1975 the importance of oil to the U.S. economy rose from 20 to 35 trillion British Thermal Units (BTU); import dependence of the United States rose from 23 percent to 39 percent. It was in this period that the United States ceased to be able to meet its energy requirements from indigenous resources and ceased to be the emergency oil supplier to Japan and NATO allies.
3. The Middle East and Africa remained the single most important source of oil for West Europe and Japan, and its importance to the United States nearly tripled from 1960 to 1975.
4. Throughout 1960-1975, only the USSR was and remained energy self-sufficient and hence had no supply vulnerability.

Sources of Oil

The locations of the immense reserves from which the world draws its oil are quickly summarized.[1]

From the beginning of the modern use of petroleum, the greatest reserves of all have been found in the Middle East; the greatest single-nation production, until recent years, came from the United States and the Soviet Union. Presently, the USSR seems to lead, with Saudi Arabia narrowing the gap (at 9+ millions of barrels per day [mmb/d]) and having no technical difficulty in surpassing the Soviet Union. But oil in

Table 1. Energy Consumption and Oil

	Energy consumption (QBTu)				Oil and gas as a percent of energy consumed				Percent of oil consumed met by imports				Middle East and Africa supplied percent of oil imports			
	1960	1965	1970	1975	1960	1965	1970	1975	1960	1965	1970	1975	1960	1965	1970	1975
United States	44	53	67	73	73	74	76	75	22	20	23	39		10	8	23
West Europe	26	34	46	50			64	71	100	100	100	100	72	83	94	95
Japan	4	6	12	14	39	60	72	78	100	100	100	100	74	82	92	77
U.S.S.R.	26	34	45	60	38	50	58	59	0	0	0	0	0	0	0	0

world trade—which is the essential point, for it provides the supplementary amount for importing countries—has been predominantly from the Middle East. Today, that oil accounts for nearly 75 percent of the oil in world trade.

Reserve figures indicate the volumes of oil in a given field, country, or region that can be extracted profitably at prevailing prices and with available technology. Producing capacity is how much can actually be extracted with existing producing facilities. The reserves of key producers (with their present producing levels and "spare capacity" indicated) and their share of world trade in oil give the fullest picture of the contemporary importance of these states. OPEC members dominate world oil; their oil production is about 30 mmb/d. Communist nations produce an added 12 mmb/d, and the rest of the world adds about 16 mmb/d. OPEC oil reserves ("estimated" and probable) also dominate: nearly 480 billion barrels, compared to 65 billion barrels for the communist states and 120 billion barrels for the rest of the world. But the nations of the Gulf proper have an importance of their own. They actually produce 23 mmb/d (and may have the capacity to produce 28 mmb/d); their reserves are presently calculated to approach 365 billion barrels.

Logistics and Refining Requirements
of the Industrial World (1977)

There are three key requirements in oil supply: production, transportation, and refining. Present production of oil destined for world trade is now, for all practical purposes, under the political control of OPEC states. (Canada is presently the only exception, but official government policy has been to phase out oil exports in favor of domestic use of indigenous oil production.) "Control" is not absolute, of course, but the producer governments' role in determining levels is unprecedented.

The transport of oil is now virtually under the control of states not party to OPEC. The overwhelming majority of tankers in the world oil trade belong or are under charter to the private international oil industry. In the event of supply shortages, these ships are still under sufficient control to be responsive either to the companies' direction or, if the companies are under extreme producing-government pressure, to the direction of the leading industrial nations (but not the Soviet Union). As for tonnage, it has proven adequate in the three major supply problems of 1956, 1967, and 1973, but it was exceptionally tight in 1970. It is more than adequate in 1977, including vessels of lesser size required for U.S. ports; the present general surplus of tonnage is about 40 percent. The Soviet Union appears to possess sufficient tanker tonnage to meet its needs and supply commitments. The states party to OPEC do not seem to have acquired more than 3 percent of the tanker tonnage in world oil trade.

Pipelines are found mostly within the United States, Canada, Europe, and the Soviet Union. They are essential ingredients in the secure and continuous supply of oil and gas. None of these is, of course, under the control of any state party to OPEC. Only the oil and gas pipelines supplying Europe with Soviet oil and gas need be considered presently vulnerable to a politically motivated closure, or for reasons of economic warfare. The pipelines of Iran and the system by which Iraq and Saudi Arabia supply oil to the Eastern Mediterranean are under the exclusive control of the host country (or shared in transit). None is essential (by reason of volumes shipped and alternative terminals and routes) to oil in world trade. The addition of the new Saudi pipeline to the Red Sea may not change this situation.

The supply vulnerabilities in the present logistic system relate to: (1) that fractional contribution to Europe's energy provided by or transitting through the Soviet Union; and (2) the future size composition and functional breakdown of the world tanker fleet. With regard to the former, the volumes of supply from the USSR are not, in 1977, important amounts by any standard, but they are obviously something to watch, especially with regard to gas. With regard to tankers, the trend toward larger size and greater tanker specialization could in the future reduce the flexibility of the logistics system.

As for refineries, there is presently no shortage in process-ing crude in any of the leading energy-consuming states. Each possesses ample refining capacity to handle its energy needs (in the case of the United States, the key Caribbean refineries are considered "secure"). The expressed intentions of the oil-producing governments to move downstream have, to date, been unfulfilled. Refineries are under the control of in-dustrial nations. In product supply, only the Caribbean re-fineries raise questions of sea-lane security; the vast bulk of petroleum in world trade moves as crude.

Moreover, in no instance do OPEC members individually represent an important segment of world oil consumption; nor collectively do their internal needs total a significant amount: some 2 mmb/d. The vital, large-volume markets for OPEC oil lie solely in the industrial nations. In summary,

		Controlled By
Function	*OPEC States*	*Industrial Nations*
Production	X	
Logistics		X
Refining		X
Markets		X

International Oil Industry

Credit for the place of oil in world energy goes to the pri-vate oil companies, especially the "international majors," whose extensive interests, management skills, application of capital and technology to oil exploration and field develop-ment, logistic systems, processing facilities, and then delivery to consumers were woven into an integrated operation of enormous influence and benefit.

From the outset, the domination of international oil by British and American companies has been a "constant"; in 1977 it is still possible to note that there are no close "seconds." Some erosion of their role has come from ENI, CFP, and other national companies of very diverse national origins. Some has also come from the scores of "independents" that entered international oil after 1950. But the direct actions of governments have inflicted the greatest changes on the international oil industry.

Quite apart from the actions of producer governments in asserting control over the disposition of their oil, the actions of consumer governments to limit the corporate freedom of the international oil industry have been long-standing, persistent, and increasingly successful in "sunshine" measures forcing disclosure of information on prices, profits, and planning. The erosion of the role of the international majors has thus derived from both the producers' and the consumers' role.

Nevertheless, certain functions exercised by the "majors" continue to be irreplaceable: the management of worldwide movements of varieties of crude, and access to tankers, refineries, and markets on the commanding scale necessary to move huge quantities of petroleum.

Their earlier role as generators of capital has diminished greatly in importance, at least within the producing nations. Even research and technology may no longer be so closely held by the majors. However, the extensive application of technology is a very complex undertaking and is still considered to be the majors' special province, as is the application of these skills to field development in enormously difficult and exacting undertakings such as the North Sea and Alaska.

The international majors have lost much of their power to determine producing volumes and prices, and they are beginning to lose the ability to make plans and commitments independent of consuming governments as well. The majors are widely regarded as being responsive or captive to policies and directions from their own governments; it is believed that in effect they are and have always been "instruments of foreign policy" (as Acheson unfortunately and inaccurately described them).

Depending on one's viewpoint, the fact that the companies have not been such is either to their credit or can be attributed to the failure of governments to define their energy interests and the parameters within which the companies would operate. In any case, the companies' role is still indispensable to all interests—albeit more limited than before in setting the economics of the oil trade.

Policies of Energy-deficient States

There have been two phases in the history of the oil policies of these states: *first* came the imperial period, in which governments and oil companies competed for oil concessions; government support for these arrangements was always thought to be the ultimate guarantee of their durability.

Of course, for the international oil companies, then as now dominated by British and American giants, priorities were reversed: their commercial interests were paramount; for them, the rivalries between states were aspects of the perennial problem of access to ever-greater volumes of oil—to be taken advantage of when necessary. The international oil companies did not encourage their governments to develop energy policies that might limit the commercial freedom of action deemed essential to their worldwide operations. Support or protection from governments? Yes. Guidance or direction? No.

The second phase—which began in the years between the two world wars—was characterized by the rise of government oil companies or chosen entities whose purposes were interrelated: (1) to provide for national involvement in the supply of a commodity becoming critical in its importance; and (2) to challenge British and American domination of oil in world trade. In the first purpose, the central concern was to extend a more effective reach over the activities of one's chief suppliers and to appraise better the terms on which oil was imported. In the second objective, governments—perhaps more to "show the flag" than to gain a commercial advantage—

encouraged international oil activities of national companies. These two purposes, often linked, have grown in importance and consequence.

The rise of consumer governments' oil companies has been nearly simultaneous with the emergence of producer governments' oil entities, thus reflecting at least a shared interest in the terms on which oil in world trade is supplied.

Involvement by governments of oil producers and importers has come to overshadow the importance of the commercial stakes; the supply of oil has become so critical a national interest that factors other than the economics of trade necessarily intrude. As a result, governments may now use all the instruments at their disposal to assure themselves of an adequate and continuous provisioning of oil at an acceptable price; from the viewpoint of producers and consumers of oil, other interests become involved in access to oil: military assistance, technology, investments, economic and political objectives, all of which complicate infinitely the context in which energy resources are traded.

In the process, national oil companies of consumer countries have begun to acquire a potential for acting directly or indirectly as instruments of policies that reflect a broader range of concerns and that are less consequential in helping set the commercial terms on which oil is supplied. The national oil companies of producer countries largely set these terms today, and these conditions may also reflect a very broad range of producer interests, of which "trade" is only one, albeit a very key one.

Against this background of the noncommercial aspects of oil, the complexities involved in the process of attempting to assure supply were all too apparent in the negotiations that led to the creation of the International Energy Agency for consuming/importing nations. Its ostensible purpose was to agree on an equitable means for sharing available oil in the event of another emergency; agreement has been reached on details, including an emergency stockpile program that eventually may provide for ninety days of imports. But from the time the U.S. government pushed hard for the IEA, the con-

suming nations were (and remain) apprehensive that if another embargo or cutoff comes, it will probably be aimed
largely at the United States and that they will all be impli-
cated by virtue of the fact that they share oil in world trade.

Behind the clear hesitation of other consumers to commit,
in advance, to such sharing was the more basic issue; should
the IEA prove to be "confrontational" in the eyes of the producers, would not Europe and Japan be risking far more
than the United States? Thanks mainly to the skill of Etienne
Davignon, who rallied the Common Market, the IEA was
brought into existence, thus raising one of the fundamental
pillars of U.S. strategy for "dealing with OPEC"—namely,
a united front of consumer nations.

Nevertheless, no IEA member forgets for long how intri-
cately interwoven are diplomacy, politics, energy, and economics, and how inadequate the IEA alone will prove to be if
much more is not brought forward to persuade the producing
world to meet the rising oil demands of the industrial world.

Such an effort began in December 1975 in Paris. The
launching of the Conference on International Economic Cooperation, the result of a Saudi and French initiative, began
the process of discussing the interrelated issues of energy,
other commodities, economic development, and financial
questions. The CIEC approach was opposed initially and, according to some, consistently by the United States, but it was
generally desired by others, who saw in it, at least, a step
away from "confrontation" and possibly toward some more
satisfactory understandings of which reliable access to energy
would be a critical part. It is another effort to foster a more
mutually satisfactory relationship between the suppliers and
consumers of raw materials than the former imperial system
provided. It is not at all certain that CIEC will succeed:
great interests are being engaged, and many of them do not
yet seem to be susceptible to compromise.

Energy and the Soviet Union

Of all the leading industrial nations, only the USSR is

presently energy self-sufficient. Soviet oil and gas production meet current domestic requirements. So far as can be determined, USSR internal energy meets an overall demand of 60 QBTU out of a supply that appears to draw upon coal for 34 percent, gas for 23 percent, and oil for 36 percent.

The principal current energy problem confronting the USSR is the increasing difficulty it has in meeting the energy requirements of the East European countries while insisting it be the principal supplier. Current East European consumption of oil is about 1.8 mmb/d; the USSR supplies 90 percent of the region's imports, mainly from its own resources, and the balance is obtained primarily from North Africa and the Persian Gulf.

The further exploitation of Soviet energy resources has entailed immense costs. Topography, logistical and technical difficulties, and the long pipeline systems required to tap East Siberian basins will continue to be impediments to development. To date, Soviet efforts to acquire the technology (drilling, offshore designs, and recovery techniques) presently possessed very largely by Western and Japanese oil companies, which is thought to be essential to further development of Soviet energy resources, have been largely unavailing. But the Soviets persist. Estimates of Soviet import requirements over the next decade vary from slightly over 1 mmb/d to 9 mmb/d. We simply do not know and can argue only that the Soviets may have to import, during the next decade, until their undeveloped additional resources can be exploited. Western technology and pipe would lessen the gap and allow the USSR to regain energy autarky. The Central Intelligence Agency report of April 1977, which estimated a level of Soviet dependence on the oil market (Middle East) of 3.5 to 4.5 mmb/d (1985), is considerably above the degree of dependence that most others forecast.

A question about Soviet policy and energy that has long haunted observers is whether the Kremlin could succeed in aligning "radical" producing governments on its side in a program of economic warfare against the free nations of the West and Japan. There have been opportunities for such efforts: the Suez Crisis of 1956, the June War of 1967, and the October War and embargo of 1973-1974. Moreover, it would seem that in all three instances there was more than enough anti-Westernism to warrant Soviet expectations of an increase in its influence. Furthermore, huge Soviet outlays for loans, development projects, and military equipment and training at various times and places since World War II— Iraq, Syria, and Egypt, now Somalia and Ethiopia—must have seemed to the Kremlin reason enough to believe it was creating a permanent presence in the Middle East, which it had long sought but which always seemed elusive. Still, none of these initiatives proved durable enough, to date, to endanger oil supply. Until now at least, the USSR's campaign against "oil imperialists" has failed.

Why it has been unsuccessful is conjecture, but three considerations seem valid:

1. The Soviets have failed to identify their own objectives and ideology with those of even "radical" regimes in the Middle East and North Africa, despite occasional and relatively short-lived tactical relationships that have not been pervasive enough to become "permanent" or "strategic." However radical many Arabs may seem, probably very few are willing to invite in another alien direction after the Arabs have just discarded that of the West.

2. The Soviets could not penetrate deeply enough in Kuwait, Iran, and Saudi Arabia, whose oil interests dwarfed the only oil-producing client state the Soviet Union had acquired: Iraq. Without strong influence in these other producers, Iraq was never enough. Iraq may not now be considered "reliable" by the Soviet Union. Moreover, Kuwait, Iran, and Saudi Arabia have been conservative forces in the region, the least inclined to stake their futures in the communist world.

3. The Soviets have never been able to persuade the producing countries that any possible disruption of oil supply to the West caused by the actions of the producing governments vis-à-vis the international oil companies

could be offset by an immediate replacement of the companies' functions from the talents available in the producing states with assistance from the Soviet Union. Finally, there was never any prospect that the communist world's purchases of oil from the Middle East would ever be of such magnitude as to provide even approximately, or even for a short time, a level of revenues that could be earned in present and foreseen Western and Japanese markets.

There may be a fourth explanation: the USSR may have judged that its efforts to disrupt oil arrangements with the West might provoke a response from the United States that, in time of crisis, could lead to general war. For that, no crisis, however genuine or Soviet-initiated, was yet worth such an outcome. Or is it possible that short of a general war, the Soviets may have been unsuccessful in their search for willing allies because the Kremlin would or could not offer assistance on the scale potential clients deemed necessary?

Policies of Producing States

As for the evolution of the policies of key producing states, the critical observation is that each key oil-exporting nation (and nearly every one of lesser consequence in the oil trade) has passed through a variety of colonial experiences under Western empires; if the country was not a colony in a formal sense, its leadership and people may have regarded themselves as colonies. Since in many cases the initial exploitation of their oil resources took place during a neocolonial experience, their assumption of national control over the disposition of their oil came at a time they regarded as marking the end of that era. Thus, for virtually all of them, "oil" has profound significance in their political and economic emancipation. The list includes Mexico, Venezuela, Algeria, Libya, Iran, Iraq, Kuwait, Indonesia, and Malaysia.

In its escape from a "colonial" relationship and in its

effort to assert national control over oil, Mexico led the way in 1938. It expropriated the holdings of foreign oil companies. That was nearly forty years ago, yet there is scarcely a more volatile political issue today than oil: Mexico will determine the disposition of its oil, not foreigners and foreign interests.

Iran—under Mossadegh—had particularly explosive implications. If national control of oil had been assumed, it would have had an incendiary effect on oil concessions and the role of major oil companies throughout the Middle East. While Mossadegh "failed," largely because of the British and U.S. oil companies' embargo, the outcome came to be regarded as, in fact, an Iranian victory. In the end, the near-monopoly of the Anglo-Iranian Oil Company was replaced by a consortium, with significant U.S. company participation, which was to function under very different circumstances. Iran's eventual acquisition of control over its oil was seen to be an essential part of economic and political independence, and the Iranian example came to have implications for oil developments in other less-developed countries (LDCs).

Indonesia was also a pacesetter. General Ibnu was convinced that as long as the colonial-style concession system prevailed, the corrosive domestic politics of oil could not be contained. He insisted that a new oil regime was essential, and the companies were forced to comply. Production-sharing/service contracts became the general rule and very largely and quickly removed the political stigma from foreign exploitation of Indonesian oil.

In the case of Nigeria, exploitation of its oil reserves began after political independence, but its disposition today is seen against the background of its colonial experience and a chaotic post-independence period, which was further complicated by the machinations of foreign powers. Nigeria's attitudes and oil policies are, therefore, based on the belief that economic independence is an essential concomitant to political independence, the same belief that motivates other LDCs.

Canada and Australia may be thought to be exceptions to

the general link between oil and "colonialism," but they are not. No one who has followed political developments in either country can be unaware of the widespread attitude that their resources are for them to develop; in both cases, there is a strong belief that their energy industry has been too long dominated by foreign interests. For example, nearly 90 percent of Canada's oil is controlled by U.S. oil companies' subsidiaries. Australia's experience has not been much different. In the matter of oil, their sensitivities are as acute as those of OPEC exporters.

Some regard Saudi Arabia as the truly significant exception to the colonial experience of other producers. Originally, Saudis preferred U.S. oil interests over the British, largely because the United States was "different" and could be a counterweight to British domination of the Persian Gulf. To date, the Saudi-American relationship has remained largely free of the antagonism that has characterized the relationships between other producers and those who controlled the disposition of their oil. Nevertheless, given the enormous importance of Saudi oil, it is necessary to consider present and future prospects in the light of three observations: first, it is entirely conceivable that at a critical moment some of the Saudi leadership will make political capital out of the incontrovertible fact that U.S. oil companies have, from the outset, monopolized the exploitation and disposition of the immense Saudi resource; second, the assumption by Saudi Arabia of political control over oil has come at the same general time as that of other OPEC states, so that the political éclat of taking national control over oil joins the Saudis to other producers; and third, the Saudis cannot isolate themselves from their setting in the Gulf. They will not be immune to political forces and trends in the area.

In summary, the warning signals of political change in the control of oil came early, and then at an accelerating rate. There was little evidence that the principal actors—the international oil giants—anticipated any change. As it was, when the storm broke, the politicization of oil spread at a rate that precluded timely adjustment on the part of the companies and their governments, even if they were disposed to attempt to do so (which they were not).

Some of the producers' dislike or apprehension of the international oil companies comes from the obvious link between British imperial interests and British control of a significant share in two of the ubiquitous "Seven Sisters." The other five "Sisters" have been American-controlled, and here the general dislike or apprehension has been directed against this symbol of "international capitalism" as far as the majority of socialist leadership in LDCs are concerned. All of the important oil-producing and oil-exporting nations are also LDCs.

Observations on Energy (1977)

1. The unavoidable dependence of virtually all major energy-deficient states upon Middle East oil is the primary fact. So great is the present underutilized production capacity of the region (about 5 mmb/d) that if all other sources of oil in world trade shut down, major industrial states might have their essential energy import requirements from that region's spare capacity alone.

2. For the past several decades, outside of the Middle East and communist world, there have been no discoveries of large oil reserves other than in Libya, Nigeria, the North Sea, Alaska, and Mexico. A conservative estimate of the time required to explore, prove, develop, and produce a truly significant amount of oil for world trade is from five to ten years. It would be exceedingly fortunate, and not to be counted on, were such discoveries to result in production that exceeded the world's increased use of petroleum and readdressed the failing reserves-to-production ratio to offer some promise of adequate supply through the remainder of this century.

3. There is no prospect, therefore, of any diminished importance for Middle East oil over the next decade and probably for years afterwards.

4. The USSR (and PRC) are not now major contenders for

Middle East oil. Competition for Middle East oil is between NATO allies and Japan.

5. Generally, OPEC is less significant today than is OAPEC. More exactly, Saudi Arabia's present spare producing capacity and potential production are most important. Saudi production is about 9 mmb/d; its present production capacity is estimated at 11.5 mmb/d; and its potential may be 20 mmb/d or even higher. Any Saudi decision on volumes and prices is important.

6. The key question of the Organization of Arab Petroleum Exporting Countries (OAPEC) is the potential divisiveness within the Gulf, especially among Iraq, Iran, Kuwait, and Saudi Arabia, and the problems and opportunities this may present to the great industrial, importing nations and the Soviet Union.

7. The United States' "special" relationships with Iran and Saudi Arabia are far and away the most significant links between producers and great industrial oil importers. Yet Europe and Japan are far more dependent on Saudi and Iranian oil than is the United States, which suggests that the Saudi and Iranian connection could prove troublesome for intraalliance relations. Furthermore, given the general hostility in Saudi-Iranian relations, U.S. "special" relationships with both may be increasingly difficult to manage over time.

8. OPEC member states—all LDCs—control the production of almost all oil in world trade.

9. Producing countries control a significant portion of the sea-lanes, which begin with their terminals and loading facilities and extend throughout the Gulf, into the Indian Ocean, the Red Sea and Suez, the Mozambique Channel, the Straits of Malacca, and Lombok. All these are susceptible to interference or closure by producers or LDCs or both.

10. Control of the rest of the logistic system—tankers, pipelines, and processing facilities (and energy technology)—still lies with the industrial world, as do the great oil markets. Only about 3 percent of the world's tanker fleet of 320 million tons is regarded as presently under the control of the OPEC states. Only pipelines that deliver oil to terminals for

tankers in their hands; none of those serving consumer nations directly is affected. The consuming nations have ample
refining capacity today and are dependent on no one else's.

11. Europe, Japan, and the Middle East have superport
facilities commensurate to their present needs. The United
States cannot take full advantage of the economics of very
large crude carriers (VLCCs). Because of its lack of deepwater
ports, the United States is crucially dependent upon smaller
vessels, whose replacement has lagged behind the construction of VLCCs.

12. Member states of the International Energy Agency,
through their plan for the sharing of available oil supplies in
the event of supply disruptions for any reason, may be able
to withstand supply interruptions or cutbacks better than at
any previous time. At least, the emergency sharing mechanism suggests that this may be so, thereby introducing uncertainty into the calculations of the oil producers.

13. International oil companies remain essential to consuming and producing nations alike, largely because of their
logistic systems and access to processing facilities, which
permit them to handle very large volumes of oil. At least 80
percent of world oil trade is handled by these international
oil companies, or some 28 mmb/d. In any twenty-four-hour
period some one billion barrels of oil are moved from one
place to another.

14. In the very years when oil became of such critical importance to the industrialized countries, "control" over access to the resource was wrested by producers from the oil
companies of consuming states. As of mid-1977, no general
arrangement between the producers and consumers of oil has
been reached providing dependable assurance of supply; there
is not even agreement on a process for doing so. Producer nations largely determine prices, and consuming governments
have not yet found the means to influence them. Arrangements regarding prices and volumes are still "concluded"
between companies and producer governments, although the
latter are key.

15. No major industrial, oil-importing state has yet had a

comprehensive, disciplined energy policy and program. The United States is no exception. Even if the oil-importing states had well-defined goals and commitments, they could in no way fundamentally alter the basic aspects of the contemporary geopolitics of energy. Very substantial efforts, persisting over a great many years, are the conditions for eventual escape from the current situation in which the industrialized states are dependent on oil resources controlled by a small group of LDCs.

2
Oil, 1977-2000

The geopolitical significance of oil derives from two central factors: (1) oil, like fuel and feedstock, is the lifeblood of the industrialized economies; and (2) oil reserves and production tend to be geographically concentrated in certain less-developed countries. In effect, oil reserves and production are most abundant in a small number of developing countries, while the need for adequate and continuous supply of oil in very large volumes is most urgent in the developed, industrial states.

The Continued Importance of Oil

Oil will continue to provide the bulk of total energy needs until 1985, and almost certainly into the 1990s. Forecasts considered in the preparation of this study indicate that oil will constitute some 50 percent of the total Free World energy supply in 1985 and not much less in 1990.

However, there is a strong possibility that the forecasts underestimate and abbreviate the importance of oil:

1. Oil is considered the "swing" fuel, compensating for all shortfalls in the production of alternative energy sources; to the extent that shortfalls do materialize in the production and development of coal, natural gas, and nuclear energy, oil will be called upon to play a

31

greater role.

2. a. The forecasts assume that, in the future, GNP growth
rates will be below historical trends; if future GNP
growth rates return to trend or exceed the forecast assumptions, oil will have to supply a greater share of increased energy requirements. Since forecasts tend to be
overly influenced by current and short-term events,
there is a reasonable chance that the assumed growth
rates are, in fact, too low.

b. To the extent that reduced oil demand was more a
result of recession than of higher prices, economic recovery should spur oil demand.

3. To the extent that the price of near-term energy alternatives has moved in line with world oil prices, the incentive for substitution (which may also involve purchase
of new equipment and other investment costs) is reduced, while the incentive for developing indigenous
sources of oil has increased, i.e., the switch from oil to
other sources may be delayed.

4. The crisis mentality that developed as a result of the
1973 embargo has receded, and there appears to be
much less sense of urgency regarding the need to develop costly alternatives to oil, at least among the general
public.

Moreover, the predicted developments may depend on the
ability of governments to elaborate, enact, and implement
comprehensive national energy policies. The inability of governments to do so to date results in slower development of
alternative energy sources than might otherwise be possible.
Since the forecasts depend on the implementation of national
conservation programs, they may overestimate the degree of
political will existing in the industrialized countries. Generally weak economic performance in the industrialized countries
has led governments to oppose the increase of domestic oil
prices to world price levels, thus dampening the conservation
and incentive effects of higher prices. In addition, the continued uncertainty in some countries about the roles of the

government and the private sector inhibits the energy activities of both. In all these cases, the result is continued growth in oil demand and less rapid development of energy alternatives.

The lead times associated with developing alternative or additional energy sources are long. The following estimated lead times (years from decision to start-up)—probably optimistic—are indicative:

- Development of proved, but nonproducing field, Middle East   1-2 years
- Production from extensions of oil fields, U.S.   1-3 years
- Offshore (U.S.) from lease to peak production   9-14 years
- Surface coal mine   2-4 years
- Underground coal mine   3-6 years
- Oil, geothermal, synthetic power plants   5 years
- Coal-fired power plant   5-8 years
- Hydroelectric dam   5-8 years
- Production of oil and gas from new fields, U.S.   3-12 years
- Uranium exploration and mining   8-10 years
- Coal gasification   10-15 years
- Tar sands and oil shale   5-10 years

The impact of very long lead times is made clearer when one adds to these figures the delays in decision making resulting from ambiguous government policy and the fact that start-up from these sources is not the point of maximum contribution.

All available evidence suggests that only after 1985, and probably closer to 1990, can alternative energy sources—shale oil, oil from tar sands, coal gasification, nuclear—begin making a significant contribution to total energy supply.

Timing, then, becomes crucial. It is entirely possible that alternative energy sources will not be available fast enough or in enough quantity to prevent sporadic energy shortages and the development of a generally tight oil supply-demand situation. It will prove difficult to coordinate the many aspects of energy supply-demand, and sporadic shortages and strains can be anticipated, beginning as early as the first years of the next decade.

The dominant place of oil in the total energy supply is secure through 1985. Indeed, the factors enumerated above make it extremely unlikely that oil will be unseated even by 1990. If remedial action is not taken promptly, the rest of the century may be very much the same: oil will have a dominant place in energy supply. Moreover, even if the share of oil in total energy supply can be reduced toward the end of the century, the volumetric demand will be greater.

Importance of Oil Imports

The industrialized countries will remain dependent on oil as the primary source of energy, and this in turn means that the industrialized countries, in common, will remain dependent on oil imports. However, there are differences among the Free World countries as to: (1) the importance of oil to the economy; (2) the degree of dependence on imports; (3) potentials for energy conservation in general and oil conservation in particular; (4) the likelihood of increased indigenous production; (5) vulnerability to oil shortages; and (6) natural resource endowments that provide energy options.

For Japan, oil will constitute the bulk (70-75 percent in 1980, and 65-70 percent in 1985) of primary energy consumption. Moreover, virtually 100 percent of Japan's oil supply will still be imported in 1985. The longer-term outlook is not favorable, and there is no possibility for the discovery of sizable domestic reserves. In the short term, Japan can hope only to diversify its sources of oil imports to reduce its current overwhelming dependence on Middle East supplies (75 percent of Japan's crude oil imports in 1975), while the creation of a very substantial strategic reserve of crude could reduce Japan's vulnerability to supply disruptions. In the longer term, only the development of alternative energy sources—particularly nuclear—will reduce Japan's dependence on oil and, hence, on oil imports. Energy dependence, however, will remain a fact of life and of gravest strategic consequence for Japan, since it has no sizable reserves of any energy resource—neither coal, nor uranium, nor probably natural gas or oil.

Europe, too, will remain heavily dependent on imports of oil. Oil will account for 50 percent of energy consumption in 1980 and 1985. The addition of North Sea oil and gas may ease the import dependence situation somewhat—particularly for Britain—but oil-import dependence of 70-85 percent is forecast for the European Community (Europe of the Nine) in 1985. In 1985, when oil will account for close to 40 percent of U.S. energy consumption, the United States may still derive 50 percent of its oil supply from imports. Fifty percent oil import dependence through the 1990s is not unlikely. The president's goal of a reduction in imports by 1985 to a goal below 6 mmb/d is probably out of reach. It is scarcely credible that in less than eight years the giant energy consumer could accomplish this objective. (Oil represents a smaller share in total energy supply for the United States than for either Europe or Japan. In addition, U.S. oil import dependence is less than for the other two areas. Finally, the United States energy resource base is far more favorable than that of either region.)

Soviet producing fields are in decline, and new potential producing areas are located in harsh physical environments, far from markets in European Russia and Eastern Europe. The Soviets, relying on their own capabilities, will be able to develop the East Siberian fields, but the process will be a long one. Western assistance at some later date cannot be ruled out, but the longer it takes to get agreement, the less essential Western assistance may be. East Siberian oil might not make any significant contribution to Soviet oil supplies before 1985, and 1990 may be a more realistic date. At that time, Siberian oil may only compensate for the exhaustion of the older fields.

The period 1976-1985, then, will see the Soviets hard pressed to fulfill the traditional goals of Soviet oil policy: additional oil supply for the conversion of the Soviet domestic economy to oil; oil supply sufficient to meet some percentage of East Europe's oil requirements; and oil supply sufficient to provide Soviet oil exports to Western Europe. It will become increasingly difficult for the Soviets to strike an acceptable balance among these goals.

Some slowdown in the conversion of the domestic economy to oil may be anticipated. Alternatively, the Soviets could prefer to seek additional oil from the Middle East. In Eastern Europe, the Soviets will encourage these countries to get some greater proportion of their oil requirements from the Middle East. The Soviets are pledged to provide 67 percent of Eastern Europe's oil needs, a decline from the recent 90 percent policy, but in quantities sufficient to maintain Soviet influence and control. The Soviets will try to maintain some level of exports to Western Europe, perhaps even diverting crude shipment from Eastern Europe to do so—only by securing hard currency will the Soviets be able to purchase the technology and equipment to develop Soviet oil resources for future Comecon supplies.

At the same time, given rapidly expanding Soviet domestic and East European oil demand and higher oil prices, Soviet exports to the West will probably stabilize at some volume below the level that would give them significant economic and political leverage over the West Europeans. Oil exports to Japan may increase somewhat, while oil exports to Western Europe of something less than 1 mmb/d are likely. In order to engage in economic blackmail, the Soviets would have to be able to bring at least some Middle East producers along with them in the undertaking. Neither Western Europe nor Japan will substitute future dependence on Russian supplies for current dependence on Middle East oil.

It is also not anticipated that Russian oil will be viewed as competition by the Middle East states. That is, strains between LDC oil-exporters and the USSR over competition for Western markets will not develop as Soviet exports level off and as the oil supply-demand situation becomes tighter in the mid-1980s.

Interestingly, the Soviet Union is believed to contain vast "undiscovered" reserves of oil. After 1990, then, the Soviets may well regain their position of overall energy independence. Well before that date, however, Comecon as a bloc will be in deficit, and Comecon countries, including the USSR, will be competing with others in the Middle East oil market.

It should be emphasized that Soviet oil shortages will be due not to depletion of their oil resource base—however much individual fields may decline—but to their anticipated difficulty in exploiting reserves in timely fashion.

In conclusion, development of indigenous alternatives, substitution, conservation, and reduced oil demand will not eliminate dependence on imported energy resources, particularly for Europe and Japan. The energy mix may change somewhat, but it is clear that dependence on imported energy sources beyond, but including, oil will be a fact of life for the industrialized states.

Importance of OPEC Oil

Given the oil-import dependence of the industrialized countries and since all forecasters reject the possibility of any huge discoveries before 1985, or even their exploitation if there were such discoveries, dependence on imported oil is tantamount to dependence on OPEC oil in general, and on OAPEC oil in particular.

The North Slope field in Alaska and the North Sea field, discovered in the late 1960s and the early 1970s, respectively, were the first major discoveries in several decades of exploration; they were also the only ones (unless the optimistic predictions for Mexico's Reforma field prove true). Moreover, in today's energy-hungry world, major finds on the order of the North Sea or Alaska are not sufficiently large to challenge the dominant position of the Middle East. Since 1960, discoveries outside the Middle East have added less to reserves than production has subtracted. It would take the discovery of several staggeringly large fields before the role of the Middle East could be challenged.

Forecasts suggest that OPEC's contribution to world oil supply in 1985 will range from 55 to 64 percent. The Middle East and North Africa may represent 43 to 54 percent of oil supply in 1985. Beyond 1985, in the absence of immense discoveries, only the Gulf states may have spare producing capacity, and the importance of Saudi Arabia will be overwhelming.

Summary

Oil will continue to provide the major portion of total energy supply, certainly through 1985, and probably into the 1990s. The industrialized countries will remain dependent on oil imports to meet the lion's share of oil demand. Dependence on imports, given little likelihood of major new oil discoveries, is equal to dependence on OPEC oil, particularly Middle East and North African oil.

Dependence on energy imports is an inescapable fact of life for Europe, Japan, and the United States. Even assuming that natural gas and nuclear energy play larger roles in energy supply, natural gas, uranium, and possibly enriched uranium will also have to be imported. However, the natural resource base of the United States—with its extensive coal resources—is more favorable than that of any of its allies.

Particular stress must be given to the complexity of, and interrelationship between, all steps in energy development and supply. Delays or inadequacies in providing any part of the infrastructure will affect the whole. Thus the large scale on which these undertakings are required may lie beyond the experience and capability of great private enterprises.

The period 1977-1985 may seem to offer an improved supply-demand situation: North Slope oil will come onstream, North Sea oil will become available in quantity, and spare producing capacity exists in the Middle East. However, if the forecasts underestimate oil's continued importance, shortage could develop even in this early period. Beyond 1985, the oil supply-demand situation will be tight, resulting in increased competition for oil from the only area likely to have spare producing capacity at that time—the Gulf states and particularly Saudi Arabia. A tight supply-demand situation shifts the balance of bargaining power to producers if there are no mitigating or countervailing factors.[1]

Oil Reserves

Concentration of Reserves

In 1976, the Middle East and Africa accounted for 65 percent of total world proved crude oil reserves (78 percent of Free World proved reserves); Middle East and African consumption, however, represented only 5 percent of world consumption. OPEC reserves constituted 66 percent of total world proved reserves (80 percent of Free World), while the narrower, all-Arab OAPEC held 50 percent (60 percent Free World). At the same time, North America and Western Europe (Japan had none) held only 15 percent of world reserves, but all three areas accounted for 65 percent of world consumption.

It is extremely unlikely that this pattern will change significantly. Instead, the trend will be toward the increasing concentration of reserves in fewer and fewer states—specifically, the Gulf states—as consuming states produce at capacity levels exceeding additions to reserves. It is not anticipated that additions to oil reserves located in the industrialized countries will exceed the growth in oil consumption. Accelerated exploration, enhanced recovery techniques, conservation efforts, and slower growth in oil consumption will not prevent a decline in the reserves-to-production ratio. Moreover, given the long lead times between discovery, development, and full production, additions to reserves now might not make a contribution to oil supply until the first years of the next decade at the earliest.

Reserve Categories

With higher oil prices, it has become easy to go beyond the sphere of proved reserves to talk about the additional reserve categories that, at higher prices, may be economic.

Proved reserves, according to the American Petroleum Institute, are those "quantities of crude oil in the ground which geological and engineering data demonstrate with reasonable certainty to be recoverable from known reservoirs under existing economic and technical operating conditions." Estimates of proved reserves are the figures to which reference is most commonly made.

Typically, the rate of recovery from a functioning field is low—perhaps 30-40 percent on average (with 20 percent as a minimum)—and additional oil is available only with the use of enhanced recovery techniques. Moreover, in the course of developing a field, reserve estimates will often be altered as a clearer picture of the field's characteristics emerges. Oil potentially recoverable from existing fields in the form of extensions to the perceived size of the field or through secondary and tertiary recovery is designated "probable" reserves. "Proved" reserves plus "probable" reserves equals total discovered reserves.

It is also possible to estimate undiscovered reserves through geological inference or the use of other sophisticated techniques. Undiscovered reserves are called "possible reserves," reflecting somewhat less certainty as to their existence and size. Combining discovered and undiscovered reserves and assuming a recovery factor of 40 percent provides a figure for "ultimately recoverable" reserves.

The total oil resource base is a measure of the total amount of oil believed to be in the earth, leaving aside the question of the economic and technical feasibility of recovery.

It is often suggested that at higher prices probable reserves should become economic and exploration for undiscovered oil should intensify. If the investment climate is hospitable, this may actually occur. But it says nothing about the constraints imposed by deficient geological-engineering knowledge, heavy capital investment requirements, availability of necessary equipment, and environmental considerations. In addition, the price will have to be high enough and reflected in the market (i.e., free of government price controls). Finally, and most crucially, the oil must actually be there to be found and developed.

Moves toward developing probable reserves and accelerated exploration will take place; the progression may not be smooth, rapid, or cost-free.

Ultimately Recoverable Reserves

World total recoverable reserves of 2 trillion barrels are thought to exist; 55 percent or 1.1 trillion barrels have already been discovered. The addition of probable reserves to proved reserves does not alter the concentration of oil resources noted earlier. Of the 1.1 trillion barrels of discovered, ultimately recoverable reserves (proved plus probable), some 513 billion are located in the Middle East alone.

Total ultimately recoverable reserves (proved plus probable plus undiscovered) in the Middle East are estimated at over 663 billion barrels. Some 513 billion have already been discovered, but relatively little has been produced, leaving huge reserves for future exploitation. The largest undiscovered reserves are believed to be in the Soviet Union and China. It is believed that out of some 478 billion barrels of recoverable reserves in the USSR, only about 178 billion have been discovered. Some 300 billion barrels, mostly in East Siberia, therefore, are thought yet to be discovered.

Large additions to reserves in the United States, Western Europe, and Japan are not anticipated. It has already been noted that "large" would not be enough in any event; only huge new fields could challenge the role of the Middle East producers. The additions to Free World reserves that will be made will come from the extension of existing fields and from offshore areas, or so contemporary opinion asserts.

Table 2. World Estimated Crude Oil Recovery,
Jan. 1, 1975 (in billions of barrels)

Region	Discovered ultimately recoverable	Expected undiscovered recoverable	Total recoverable
Russia, China, et al.	178	300	478
North America	173	155	328
United States	(157)	(85)	(242)
Western Europe	27	57	84
Middle East	513	150	663
Africa	89	71	160
South America	84	82	166
Far East	41	90	131
Antarctic		20	20
Total	1,105	925	2,030

Source: "World Oil," September 1975, p. 49 (based on article by John D. Moody and Robert W. Esser) in Congressional Research Service report, p. 41.

Oil Production

Reserves set an outer limit on what can be done; but reserve figures alone say little about what actually will be done. Clearly, different levels of reserves sustain different levels of production depending on demand, price, availability of logistic supports for exports, geological characteristics of the producing areas, technological capability, conservation considerations, and the political and economic objectives of the producing government. For example, the United States, with some 31 billion barrels of proved and probable reserves, produced some 3 billion barrels a year, while Iraq, with 34 billion barrels of proved and probable reserves, produced only 760 million barrels in 1976. The essential point here is that the intensity with which any given quantity of reserves will be exploited will be determined by a host of factors, some economic and some political.

Pattern of Production

There is, however, a positive correlation between reserves and production. Therefore, it should not be surprising to find production concentrated in areas outside the industrialized states. The Middle East accounts for 38 percent of total world production; Africa, 10 percent; Latin America (including the Caribbean), 8 percent; the Far East, 4 percent; and the communist world, 22 percent (all 1976). Only 18 percent of total world production originates in North America and Europe, and virtually none originates in Japan. There is little likelihood that this pattern will be altered.

In the United States, the addition of Alaskan and offshore production will reverse the declining trend in U.S. production for the 1977-1985 period. (In 1973 the United States produced 9.2 mmb/d; in February 1977, production was 8 mmb/d—a decline of about 12 percent). Beyond 1985, production may well decline again, although the decline will be from the higher levels attained by then. European production will increase for the next several years as the existing North Sea fields are developed and brought to full production. Beyond 1985, production will level off, perhaps until 1990, before it declines—unless one accepts the extreme estimates of what the North Sea may yet possess in recoverable oil. In any event, just as reserve additions fall below production levels (drawing down reserves), in the same way production levels will lag behind growth in consumption, and increased production will not obviate the need for substantial oil imports.

With the industrialized countries producing at capacity, production in these countries may peak and begin to decline some time after 1985. Japanese indigenous production will remain totally insignificant throughout the remainder of the century, unless offshore discoveries affect this otherwise grim outlook. The rate of increase in Soviet oil production may be slow until East Siberia is brought on-stream, an event not likely before 1985 at the very earliest. At that time, Siberian production may not add much to Soviet production but merely compensate for declines in the older fields.

The concentration of significant oil production in a small group of nations will be intensified throughout the remainder of the century. The developed countries' share in total production fell from 29 percent in 1965 to 18 percent in 1976.

By contrast, Soviet-bloc production increased from 18 percent of the world total in 1965 to 20 percent in 1976. The
comparable figures for the Middle East and Africa are 35 percent in 1965 and 43 percent in 1976. As production peaks in
other areas and begins to decline, Middle East and African
production will represent an ever-larger proportion of world
production.

Secondary and Tertiary Recovery

Great reliance is now put on the additional oil to be recovered from the wider use of techniques that allow substantial increases in field exploitation. Most estimates of the oil
in place that can be produced from primary-unaided-recovery
efforts are about 30-40 percent for known U.S. fields. U.S.
experience in these techniques is probably greater than anyone else's. Discussion of their value, therefore, is primarily
limited to the United States, although especially Iran and
Saudi Arabia are accumulating experience.

According to one major company's estimate—and it does
not differ significantly from that of other sources—of the
"attainable" potential for recoverable oil in the United
States, there are some 252 billion barrels, with 106 billion
barrels having been produced through 1974.[2] The rest, 146
billion barrels, is "available" assuming that production and
recovery techniques, and the economics, justify the effort.
It is guessed that fully effective recovery techniques could increase the percent of recovered oil from 20 percent to 37-47
percent. Future discoveries may be exploited to some 32 percent, the lesser percent reflecting an assessment that future
fields will probably lie in regions offshore or in smaller, deeper, and lower quality reservoirs onshore—fields more difficult
to reach and more costly to tap.

Thus enhanced recovery techniques are highly significant.
However, what is not understood is that secondary recovery
techniques—the use of water, steam, gas, chemicals pumped

back into a reservoir to send oil to the well—have been successfully applied to comparatively few fields and only when the field's characteristics are fully determined and correctly employed; it is a very great skill matched to a complex and sophisticated field "management" endeavor. It is not a common undertaking applicable to all or even perhaps most fields.

The vaunted "tertiary" recovery techniques, which use more advanced technology to exploit a source still further, have not really been employed outside of laboratories; they may not be available on any significant scale for another decade at least. No meaningful figure can be given as to how many additional barrels can be obtained from the use of such techniques. At a guess, it is possible that if all these techniques were successfully employed, the oil produced from existing U.S. fields could approach 65 billion barrels, 25 of those coming perhaps from the successful application of secondary and tertiary recovery techniques. It is not possible to estimate the percentage of recovery potential for the great majority of overseas fields.

Oil Shale and Tar Sands

The quantities of oil found in oil shale deposits and tar sands are believed to exceed by far current estimates of total world resources of conventional petroleum. In addition, oil shale and tar sands deposits, as far as is now known, are concentrated in the Western Hemisphere: Canada, the United States, and Venezuela. In spite of quantity and location, development of these alternative sources of oil has been slow. Moreover, in spite of the increased price of oil and the resultant greater economic attractiveness of oil from shale and tar sands, it is not expected that either of these sources will make a significant contribution to world energy supply before the 1990s, although their successful exploitation could have a very profound effect, prolonging the world's use of oil.

Oil From Shale

With regard to oil shale, the U.S. Geological Survey estimates that total worldwide shale oil resources could amount
to 161 trillion barrels of crude oil. However, this figure represents the total world oil shale resource base without regard
to the economic and technological feasibility of production.
Of the relatively accessible oil shale deposits, the U.S. Green
River deposits may include 560 billion barrels of oil in
higher-grade shale (twenty-five gallons per ton shale) and
about one trillion barrels in lower-grade deposits (fifteen to
twenty-five gallons per ton).

Oil from shale has been produced in Scotland, China,
Australia, South Africa, and the USSR. In the United States,
oil has been produced from shale only in experimental runs
and in one large pilot operation. Plans for two commercial-
scale operations of 50,000 b/d each are in the process of implementation and were to be on-stream in 1977 and 1979,
respectively. Seventy-two percent of U.S. oil shale lands,
containing nearly 80 pecent of the potential oil, is federally
owned. Recent sales have interested the oil industry, but the
oil potential of tracts leased so far is small. Expected production from these areas by 1985 is estimated at 300,000 to
500,000 b/d, or 1-2 percent of U.S. oil consumption at that
time.

Given the sheer quantity of oil potentially available from
oil shale deposits and their concentration in the United
States, what factors inhibit the greater production of oil from
shale? The technology is not firmly established, and the
financial incentives appear to be inadequate to stimulate
more rapid research and development. Of the two processes
now used for producing shale oil, the process involving mining followed by surface processing is further advanced but
far from proved; in situ processing has been demonstrated on
an even smaller scale than the first method.

Beyond the technological limitations of oil shale development, there are also serious doubts about the availability of
mining personnel and equipment. The production of 1 mil-

lion barrels a day of oil from oil shale by the surface process requires the daily mining of 570 million tons of oil shale. This mining of shale would approach, on an annual basis, the 1976 level of U.S. coal production and, when combined with anticipated growth in the coal and uranium mining industries, is beyond the capacity of available labor and capital resources.

The environmental problems associated with such large-scale mining are obvious. Surface disruption, pollution from refining, and the disposal of waste rock remain unresolved impediments to further oil-from-shale development. In addition, surface processing puts extreme demands on water resources. It is estimated that in order to develop an oil shale industry of 3-5 million barrels per day in the Green River area, essentially all available water in the region would have to be devoted to the oil shale industry. For all these reasons, oil from shale is not expected to make a significant contribution to energy supply before the late 1990s or the early years of the twenty-first century.

Oil from Tar Sands

The promise of oil from tar sands is almost as great as oil from shale, but it too will probably not be available commercially before the late 1990s. Total resources of oil in the Canadian (Alberta) tar sands are estimated at 280-560 billion barrels, and further tar sand oil resources are believed to exist in the Canadian Arctic and the Cold Lake area. U.S. tar sands might contain some 28 billion barrels (almost equivalent to current estimates of proved U.S. reserves). Venezuela's Orinoco deposits may be the largest single source, about 655 billion barrels of oil, one-tenth of which is recoverable with present technology. Data on tar sands are incomplete, but it is clear that they represent a major part of world petroleum resources.

A commercial-scale tar sands plant owned by the Great Canadian Oil Sands Company (GCOS) has been in operation in

Athabasca for a number of years. Forty-five thousand barrels of crude per day are being produced from the Athabasca tar sands. It is anticipated that production will soon increase to 65,000 barrels of crude per day. By 1985, oil produced from tar sands is expected to total between 700,000 and 800,000 b/d without any determined government support. Its prospective price per barrel—competitiveness with conventional oil—has risen sharply in the last several years; government support or subsidization will probably be required.

Shell recently canceled plans to participate in the development of the Athabasca tar sands, largely because the terms demanded by the Canadian government appeared exorbitant. Atlantic Richfield also pulled out of a project in Canada because of rapidly rising costs. With regard to Venezuela, there are serious political impediments as well as technological difficulties to tar belt development; having only recently nationalized oil company operations, the government will find it difficult to negotiate agreements with the very same companies for development of the Orinoco tar belt.

Reserve-to-Production Ratio

The concept of reserve-production ratio is designed to lend insight into the longevity of oil reserves at prevailing production levels. In actual fact, production levels are not usually constant year after year, and reserve estimates change as actual development reveals additional information about the characteristics of the field. In addition, the utility of the concept is suspect since, to date, no government has been able or willing to define an optimum reserve-production ratio. In effect, the question, "how many years of potential production at what levels is adequate?" has never been resolved. In 1976, the total world reserve-production ratio was twenty-nine years.

The Congressional Research Service report, "Toward Project Interdependence: Energy in the Coming Decade," contains an interesting calculation. If Free World oil production increases at 4 percent a year, and if oil demand increases at 4 percent a year, 844 billion barrels of recoverable reserves will be required in 1985 to maintain a reserve-production ratio of thirty-five years. Subtracting cumulative production through 1985 from recoverable reserves demonstrates that 490 billion barrels will have to be added to reserves by that date. By way of comparison, total world oil production between 1918 and 1973 was just below 300 billion barrels, and the rate of discovery of new oil reserves has averaged only 15-20 billion barrels a year since the 1940s, a figure that includes the enormous fields of the Middle East and USSR.

If, as forecast, the required future rate of worldwide discoveries is only 15-20 billion barrels a year, annual production will exceed discoveries by increasing amounts; this will draw down reserves. If one excludes, for reasons of security, reserves discovered in the Middle East and the Soviet bloc, the world's questionable ability to obtain sufficient reserves of greater reliability is evident: in the period 1950-1973 only some 105 billion barrels of proved reserves were found outside the USSR and the Middle East, or an annual rate of 4.5 billion barrels.

On the other hand, the estimates of undiscovered reserves suggest that it is technically feasible that reserves exceeding the last fifty-five years of production could be added to recoverable reserves. But progress is likely to be slow: additional reserves will be costly, and new reserves will be located in harsh physical environments. Only a fraction of additional reserves would be located in the industrialized states in any event. In addition, the issue of timing makes it unlikely that additions to reserves will be made and developed in time to meet oil demand increases. Recalling the tentative utility of the reserve-production ratio concept, the implication is still that shortages will develop in the period beyond 1985 and that scarcity will be a fact of life by the turn of the century.

Particular countries will not approach a thirty-five-year reserve-production ratio and will even fail to maintain their current reserve-production ratio. U.S. reserves represent perhaps ten years of production at current levels of produc-

tion. It is not anticipated that the United States will maintain
this ratio; rather, a deterioration will occur, since production
will increase with Alaska on-stream and since additions to
reserves will be slow in materializing. In Europe, the current
reserve-production ratio is misleading: North Sea reserves are
included, while North Sea production is not completely available. The Soviets will increase production and reserves, but
they may not maintain their current reserve-production ratio
twenty years.

It is not profitable to analyze the reserve-production ratio
more deeply. Rather, what is important is that additions to
reserves will probably be neither adequate nor timely enough
to maintain prevailing reserve-production ratios. Moreover,
the industrialized states will suffer the most rapid deterioration. Beyond 1985 Middle East spare capacity will come
under pressure, since alternative sources will fail to make
major contributions to energy supply.

Oil Consumption and Demand

The advantages of oil as a fuel source include: (1) availability in sufficient quantity at, until recently, low cost; (2) ease
of transportation; and (3) versatility and easy substitution for
other energy sources. World oil consumption quintupled over
the past twenty-five years. U.S. consumption almost tripled
from 6 mmb/d to 17 mmb/d in the same period, while oil
consumption in the communist countries increased by a
factor of ten. Japan's oil consumption was twenty-five times
higher in 1974 than in 1950, and the same time period witnessed a fourteenfold increase in West European oil consumption. In the past, the growth in oil consumption has exceeded
the overall energy consumption growth rate (Free World
energy supply grew at a rate of 5.4 percent a year in the
period 1965-1973; oil supply in the same period grew at a
rate of 7.4 percent a year).

All forecasts suggest that slower economic growth will
result in a slower rate of growth in energy consumption.

Table 3. Free World Oil Demand, Indigenous Supply
and Imports to Balance (MMB/D)

Source	Demand		Production*		Implied imports to balance	
	1980	1985	1980	1985	1980	1985
Exxon 1977	59	68	21	26	38	42
Eberstadt & Co.						
Base.	52	61	22	24	30	37
High	56	68	22	24	34	44
Congressional Research						
Service	56-58	62-68	23	28	33-35	34-40
OECD** 1977.	46	42-55	16	18-20	30	24-35
Other.	. . . .	56-61	. . . .	22-25	. . . . . .	34-36

*Production excludes OPEC production.
**OECD countries only, net import figures.

Higher prices, in addition to slower rates of GNP growth, will result in some moderation in the oil consumption growth rate. However, even at slower rates of economic growth and with higher oil prices, oil consumption will continue to increase, and the absolute level of oil consumption will remain high and require substantial oil imports. The situation may not be as unfavorable as it might have been in the absence of slower energy growth rates and higher oil prices, but oil demand and oil-import dependence remain high.

Table 3, taken from various forecasts, suggests the demand for oil over time, the likely level of non-OPEC supply, and the implied level of imports needed to balance Free World demand and supply.

The forecasts all paint a similar picture; increasing volumes of oil will be in demand, requiring increasing levels of oil imports. If developments in alternative energy sources included in the forecasts fail to materialize, the volume of oil demanded will be greater than indicated in the table. More rapid economic growth than that assumed in the forecasts would have a similar effect.

Beyond 1985, probably nearer to 1990, oil demand growth may continue to moderate as conservation efforts take effect and as alternative energy sources begin to make a contribution to total energy supply. At the same time, how-

ever, oil production in the industrialized states may peak and
level off or even decline, suggesting a continued need for oil
imports.

Oil imports will increasingly have to come from OPEC
countries, particularly from the states of the Gulf. Competition for Gulf oil will be intense. Even assuming that oil does
come to meet a smaller percentage of the energy budget, the
volume of oil required to meet the world's growing energy
demand will be even greater in the future.

OPEC

Producing capacity defines the limits of production at a
particular point in time. On the basis of reserves alone, it
seems likely that future OPEC production could meet the
needs of the world energy supply. The question is whether
OPEC will have sufficient, installed capacity to meet that
part of world demand not satisfied by production elsewhere
in the world. If that capacity exists, the next question is:
what factors—economic and political—will determine actual
production levels in OPEC countries?

Producing Capacity

Considerable excess producing capacity is currently available in OPEC. However, as world oil demand has begun to
pick up, spare capacity is already declining.

In spite of the decline in spare producing capacity within
OPEC, it is generally accepted that existing OPEC capacity
can meet world demand for OPEC oil until 1985; only after
1985 are additions to producing capacity considered essential
to meet world demand. Because of the different causes that
result in "spare" capacity, it may be that the only real
"spare" was the difference between Saudi Arabia's self-
imposed ceiling of 8.5 mmb/d (pre-December 1976) and its
capacity to produce some 12 mmb/d.

Table 4. Canadian, Mexican, and OPEC Production (MBD)*

Country	September '73 pre-embargo production	December '73 embargo production	Arab production shortfall due to cutbacks December '73	Percent	Installed production capacity	Mar. '77 production	Mar. '77 shut-in	Percentage shut-in
Algeria	1,000	820	280	25	1,000	1,000	0	0
Iraq	2,167	2,150	100	5	3,000	2,300	700	23.3
Kuwait**	3,520	2,470	1,190	31	3,500	1,900	1,600	45.7
Libya	2,286	1,770	590	26	2,500	2,210	290	11.6
Qatar	608	460	190	29	700	440	260	37.1
Saudi Arabia**	8,574	6,700	3,000	32	11,500	9,880	1,620	14.1
UAE	1,654	1,310	460	31	2,380	2,030	350	14.7
Ecuador	210	220			225	150	75	33.3
Gabon	161 (capacity)	N/A			250	220	30	12.0
Indonesia	1,338	1,500			1,800	1,720	80	4.4
Iran	5,793	6,100			6,700	6,280	420	6.3
Nigeria	2,100	2,250			2,300	2,250	50	2.2
Venezuela	3,387	3,400			2,600	2,360	240	9.2
Canada	1,745	1,800			1,800	1,338	462	25.7
Mexico	470	500			1,000	980	20	2.0
Arab OPEC	19,909	15,680	5,910	27	24,580	19,760	4,820	19.6
Non-Arab OPEC	12,989	13,470			13,875	12,980	895	6.4
Total OPEC	32,989	29,150			38,355	32,740	5,615	14.6
Canada and Mexico	2,215	2,300			2,800	2,318	482	17.2
Total	35,113	31,450			41,255	35,058	6,197	15.0

*International Energy Office, Federal Energy Administration, Washington, D.C. May 4, 1977

**Includes one-half Neutral Zone, which amounted to approximately 380 MBD in March.

Source: E. K. Bauer/P. L. Pohl; 5/477

Table 5. Free World OPEC Oil Import Requirements (MBD)

	1980	1985
Exxon	37	40
Eberstadt:		
Base	30	37
High	34	44
Congressional Research Service	33–35	34–40
Other	---	34–36

It is only speculative, but the technical limitations of Saudi fields might have imposed an additional restriction, namely, that the Saudis could not within a period of six to twelve months actually reach 12 mmb/d. It is impossible to be more specific; expert appraisals differ widely, but this may explain why Saudi production, while increasing following the December 1976 Saudi decision to increase production to keep a lid on prices, has not increased more rapidly.

At the least optimistic, some argue that *available* spare capacity today for all of OPEC may not exceed 1.5 mmb/d. If this is the case, then the Saudi decision to increase permitted production represented the only way in which world demand could be met.

Given forecasts of Free World import demand of 24-44 mmb/d in 1985, OPEC may have adequate producing capacity to meet 1985 import demand, at least at the lower levels of possible demand. The Gulf will represent well over 50 percent of total OPEC producing capacity in 1985, and OAPEC capacity will account for over 60 percent of 1985 OPEC productive capacity. The dominant position of Saudi Arabia should also be noted; Saudi capacity alone will represent 30-40 percent of OPEC productive capacity in 1985.

Producing capacity alone will not determine actual producing levels. Possible producing levels, given producing government economic needs and goals, are suggested below. These figures are based on indications received from statements of producing government policy, i.e., Kuwait has determined to restrict production levels to 2.5 mmb/d for conservation reasons. Venezuela has also limited production for conservation reasons. Iran and Iraq are committed to expanding production to 8 mmb/d and 5 mmb/d, respectively, by

1985. Other figures also reflect anticipated production levels.

In general, the trouble with the forecasts is that OPEC production is considered a residual. In other words, Free World demand is calculated, and Free World production is calculated; it is then assumed that OPEC production will inevitably cover the difference. There seems to be no inherent reason why this should be the case.

Forecasts of Free World OPEC oil import requirements are presented in table 5.

The suggestion is that the oil demand-supply situation will be increasingly tight and that, moreover, there is no inherent reason why OPEC production should attain the forecast levels. Competition for available supplies will be intense, and prices will rise. OPEC oil production may barely cover Free World oil import requirements in 1985. In 1985, only Saudi Arabia, Kuwait, and Venezuela may have spare capacity. Beyond 1985, decreases in oil demand attributable to greater contributions from alternative energy sources and higher prices may not be adequate or timely enough to prevent serious shortages.

Indeed, the possibility of shortages begins in 1980 and becomes more acute by 1985. Beyond 1985, world oil production may peak, and serious shortages can be averted only with the timely contributions of alternative energy sources.

Determinants of Supply

If OPEC production levels are barely adequate to meet Free World oil import requirements, it seems reasonable to ask: (1) what factors will encourage or discourage OPEC production at even these barely adequate levels; and (2) what factors might induce higher or lower production levels? In addition, the answer to these questions may lend some insight into future oil prices. Finally, it should be remarked that a combination of economic and political factors will determine actual OPEC production levels. In effect, given what can be done, what will be done?

Table 6. Possible Production Levels, Middle East and Other OPEC Nations, 1980 and 1985 (MMB/D)

	1980	1985
Saudi Arabia	7.9	15.0*
Kuwait	2.5	2.5
Iraq	4.0	5.0
UAE	2.5	3.0
Qatar	0.5	0.5
Iran	8.0	8.0
Total: Gulf	25.4	34.0
Libya	1.7	1.7
Algeria	0.7	0.7
Total: OAPEC (minus Iran)	19.8	28.4
Indonesia	2.0	2.5
Nigeria	2.5	3.0
Gabon	0.3	0.3
Venezuela	2.0	2.0
Ecuador	0.7	0.7
Total: OPEC	35.3	44.9

*The original figures were 9-11 MMB/D, but the authors of *Geopolitics* amended to 15 MMB/D as being more reasonable.

Source: Derived from Congressional Research Service, "Towards Project Interdependence," p. 55.

Other forecasts of actual OPEC production are similar:

	1980	1985
Exxon	38	42
OECD* 1977	34	29-39
Other	—	27-35

*Includes production for OPEC internal use and Common Market

Factors that help determine production levels include:

1. *Population.* Assuming that countries with larger populations need higher revenues to provide some minimum of
   economic and social investment necessary to maintain
   political stability and encourage economic development
   and self-sustaining economic growth, high-population
   countries need to maximize oil revenue. While population undoubtedly has some effect on the need for oil
   revenues, the goals of the government provide a more
   direct link to revenue needs than pure population statistics. The more ambitious the government's goals regarding the future of the country, the higher the revenue
   needs.
2. *Structure of the Economy.* To the extent that ambitious
   government goals can be underwritten by income from
   various sources, the need for oil revenues may be less
   compelling. While it is not anticipated that any producing government will want to settle for less than its
   subjective judgment as to an equitable return for its oil,
   additional sources of income may permit a country to
   produce less oil currently so as to extend the life of its
   reserves. Oil as a percent of Gross National Product, oil
   as a percent of government revenue, and oil as a percent
   of exports will give some indication of the importance of
   oil in any given economy.
3. *Development Plans.* Government development plans,
   even though in many of the OPEC countries planned al-
   locations are rarely actually spent, give some indication
   of the directions a country's leaders would like to take
   and the price they are willing to pay. Development plans
   will suggest the need for imports and hence the income
   needed to pay for imports. They will suggest the possibilities or limitations on the development of other sources of revenue. Development plans, then, can be useful
   indicators of a country's future income requirements.
4. *Oil Reserves.* Governments with lower reserves may be
   more cautious in their allowable levels of production.

To extend the life of reserves, conservative production levels might be adopted. Presumably countries with larger reserves can produce at higher levels and still be sure of future production. By the same token a country with low reserves but with considerable potential for developing alternative sources of revenue may elect to produce at higher levels to finance maximum development of promising economic sectors.

5. *Price.* In addition, price becomes an important consideration. If prices are high, countries with low oil reserves may be better able to obtain a high rate of income from lower levels of production, thus freeing the government from having to choose between conservation and needed revenues. The structure of the oil market will also be important. In a market with little spare capacity and a tight oil supply-demand situation, the need for deliberate production restraint is needed.

6. *Regional Politics.* The December 1976 Saudi decision to break with most other OPEC members and to institute a 5 percent price increase for 1977, as opposed to the wider OPEC program of a 10 percent increase with an additional 5 percent increase planned for July 1977, was to be bolstered by increased Saudi production. The Saudi decision suggests just one instance where regional politics may have influenced production levels. The Saudi move to increase production to put downward pressure on price may have been motivated, in part, by a Saudi unwillingness to see the Shah of Iran increase his oil revenues and thereby his purchases of military equipment; it may also have been to teach the world that the Saudis, not the Shah, are the key in oil matters. In addition, the Saudis made clear their decision was contingent upon "progress" toward a Middle East settlement (Israel) and progress in the North-South dialogue in Paris. Political factors, then, may influence oil production levels in the future.

None of the factors is deterministic, that is, while it may

seem rational to an outsider that a single factor may influence production in a particular way, it is conceivable that from a different perspective the same factor suggests a different course of action. Moreover, the factors may pull in opposite directions. In effect, all the factors suggested do not necessarily point in a single, unambiguous direction. In addition, the factors are not independent of each other, and a complex mix of these factors and others will determine actual production levels.

It is precisely a complex mix of economic factors and political objectives that will determine the now famous absorptive capacity, which, in turn, is supposed to determine production levels. A definition of absorptive capacity cannot be separated from the goals of statesmen—from leaders' visions of the future economic and political structure and role of their nation, domestically and internationally, including the emphasis placed on military expenditures. It will never suffice to say that a particular nation cannot make use of the funds generated by oil production in domestic economic development because the use of funds is tied to certain horizons and images in the minds of statesmen. Additional income itself will broaden these horizons. Increased international influence will be accompanied by additional external goals, greater opportunities for adventure, and increased international responsibilities.

Perhaps the greatest of these responsibilities derives from the very importance of oil, not merely to these LDCs, but to the industrialized nations as well. Production levels need to reflect a careful consideration of the world need for oil and the certain dependence of the Free World on oil imports. The desire for maximum oil income must not overstep some invisible, but nevertheless real, line where the threat of oil shortages, or oil prices so high as to result in shortage because of a real inability to pay, exceeds any possible cost of action by the industrialized states against the "irresponsible" oil producers.

The issue of international responsibility is a real one. It is inconceivable that the Western world could allow itself to be brought to its knees by the deliberate inaccessibility of oil. To date, even the 1973 embargo did not force the West against the proverbial wall, where the only escape would be through force of arms. A too careful balancing of producers' and consumers' needs is fraught with the danger of miscalculation. But this very task of balancing is now the job of oil producers, who will have to set production levels with one eye fixed on their own needs and the other fixed on the needs of the industrialized countries. Since their own needs include survival of the state and the regime, the need to supply adequate oil supplies is a means of insuring these.

Given the outlook for a tight oil supply-demand situation after 1985, it may be difficult to convince the oil importers that the oil producers are doing their best; competition and conflict are inherent in this situation.

Refining

By far most of the trade in oil is traffic in crude oil. Exports and imports of oil products are much less significant in world trade, reaffirming that most nations have opted for refining self-sufficiency. Products represent only some 15 percent of total oil trade. In 1976, oil products accounted for 20 percent of U.S. oil imports (down from 25 percent in 1975), originating largely in the Caribbean, and the United States accounted for over 40 percent of world oil products trade. Other industrialized states have elected a far greater degree of products self-sufficiency. The last full year of data makes this point.

Refining facilities have usually been located close to markets rather than at the source of production, thus minimizing the need for specialized products tankers. With the development of extensive refining capacity, consuming governments captured the value added, and some countries even offset the price of crude imports with products exports. Most recently, developing countries in general, and oil producers in particular, have indicated their determination to capture

Table 7. Imports and Exports, 1975 (in thousands of barrels daily)[1]

Country/area	Imports		Exports	
	Crude	Products	Crude	Products
United States	4, 105	1, 920	5	205
Canada	815	35	600	200
Latin America	2, 040	300	1, 135	2, 040
Western Europe	11, 680	930	60	185
Middle East	140	110	17, 680	825
North Africa	85	90	2, 350	55
West Africa	5	45	1, 960	15
East and South Africa	340	145		35
South Asia	295	95		5
South East Asia	1, 155	410	1, 175	280
Japan	4, 565	380		5
Australasia	225	120		50
U.S.S.R., Eastern Europe	235	70	720	750
China				
Total	25, 685	4, 650	25, 685	4, 650

[1] Includes quantities in transit, transit losses, minor movements not otherwise shown, military use, etc.

Source: British Petroleum, "Statistical Review of the World Oil Industry," 1975.

more of the benefits of their natural resources by insisting on local processing of raw materials. The intention is to replace crude exports with the export of the more valuable oil products. Trade in products would then be an increasing proportion of world oil trade.

Demand for Refined Products

Demand for refined products is concentrated in the industrialized countries. North America, Western Europe, and Japan accounted for 80 percent of total Free World demand for refined petroleum products in 1975 (a total of 36,025 thousand barrels a day [mb/d] for the three areas). In the same year, South America, the Middle East, Africa, the Far East, and Oceania together accounted for only 20 percent. Questions regarding refined products have less to do with the level or growth in demand and more with how that demand will be satisfied and from where.

Table 8. Refining Capacity, 1976

	Thousand barrels daily	Percentage of total
United States	15,237	21
Other Western Hemisphere.	9,713	14
Total North America	24,950	35
Europe	19,972	28
Middle East	3,285	5
Asia.	9,868	14
Africa.	1,328	2
Sino-Soviet Bloc.	12,406	17
Total	71,809	100*

*Does not add due to rounding

Source: American Petroleum Institute, *Basic Petroleum Data Book*

Refining Capacity

Current refining capacity is concentrated in the developed countries.

Every area, save the United States, retains product self-sufficiency. If the Caribbean refineries are included, for which the United States has been the market tributary, the United States has virtual refining self-sufficiency and capacity to spare. Moreover, there is considerable spare capacity in every area. In the first half of 1976, U.S. refineries ran at about 85 percent of their capacity; Caribbean refineries ran at 50 percent of capacity. European refineries ran at 60 percent of capacity and Japanese refineries at 80 percent. Worldwide refineries operated at about 75 percent of capacity.

Plans for Expansion of OPEC Refining Facilities

U.S. government figures for Middle East refining capacity are slightly different from those reported above. A continu-

63

Figure 1. Preponderance of Refining Capacity Outside OPEC States (adapted from International Petroleum Encyclopedia, 1975).

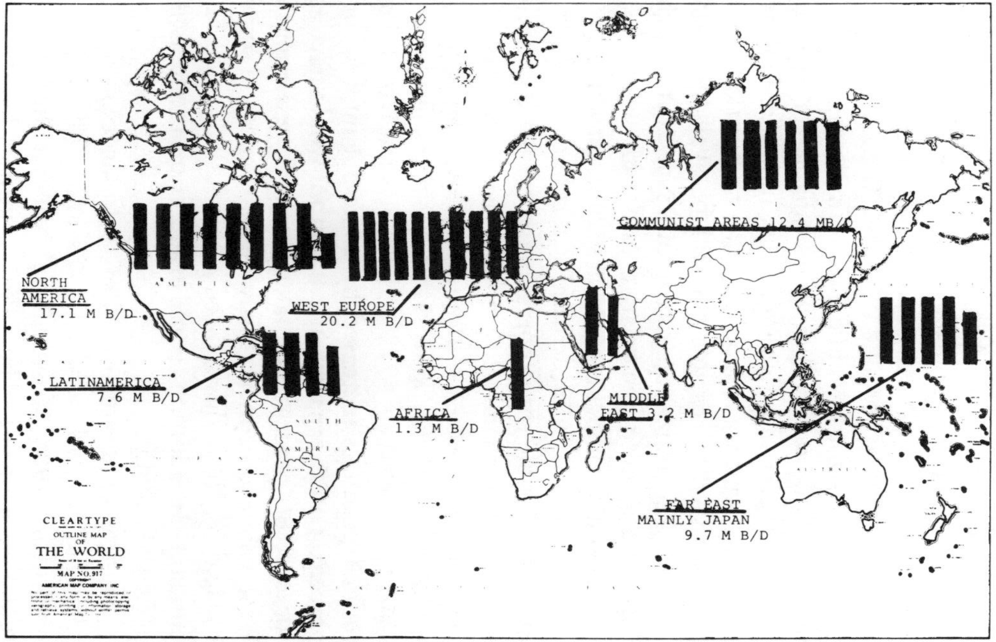

ing problem with an analysis of this type is that the statistics differ from source to source. However, the data are not sufficiently different, in this instance, to invalidate certain propositions.

The Federal Energy Administration (FEA) estimates total Middle East refining capacity at 2.5 mmb/d in 1974. Of this amount, 900 mb/d was devoted to internal demand, leaving 1.6 mmb/d of exportable capacity. The same study indicates that in 1978 total Middle East refining capacity will be 2.8 mmb/d, with some 1.8 million available for export. If some of the less certain refinery projects announced in the wake of the 1973 embargo should come to fruition, another 1.1 mmb/d of OPEC capacity would be available in 1978. The FEA used a utilization factor of 93 percent, which appears appropriate given the relative ease with which the oil producers could compel the purchase of oil products.

In terms of 1973 demand for refined products, the 2.8 mmb/d available for export represents only 6 percent of total Free World demand; 2.8 mmb/d also represents only 4 percent of 1975 total world refining capacity. However, it represents 60 percent of world trade in oil products in 1975. Moreover, between 1972 and 1973, Free World demand for refined products increased by some 3.4 mmb/d, or about 8 percent; OPEC capacity of 2.8 mmb/d would have accounted for over 80 percent of that growth. Excess capacity in the West will become increasingly burdensome, and some refining capacity in the industrialized countries may well become redundant. In addition, world trade in oil products is in for a period of rapid growth.

Further extension of OPEC refining capacity could add some 1.2 mmb/d between 1978 and 1980 and another 4.6 mmb/d by 1985. It is doubtful whether these levels will actually be achieved. The OPEC countries have been slow in getting underway with the grandiose refinery projects proposed in the euphoria caused by higher oil prices in 1973. The absolute number may be wrong, it may take longer than currently anticipated to bring these projects to fruition, but the trend appears certain.

The special security implications of increasing product dependence (as opposed to dependence on crude) is evident. A crude shortfall can be made up by drawing on other sources. But a product shortfall may not have a comparable alternative source, depending on OPEC states' policies, which may require use of their refineries, or depending on the extent to which there is adequate surplus capacity in export refineries (located elsewhere) to meet the shortage. Moreover, product shipments generally require specialized tankers, which make up only a small share of the world's fleet; there may not be enough to provide product supply from refinery locations more distant than the original refineries whose shortfall caused the difficulty.

The World Tanker Fleet
and the Logistics of Supply

The Free World tanker fleet bears the major responsibility for the efficient transport of oil from producing areas to the consuming centers. Of the nearly 32 mmb/d of crude oil and the 5 mmb/d of petroleum products traded internationally, approximately 95 percent was moved, at some point, by tanker. The adequacy, ownership, and control of the fleet are, therefore, essential elements in the geopolitics of energy.

The adequacy of the fleet refers to the capacity to move oil in the required amounts. In addition, the fleet can be assessed in terms of its ability to transport other energy sources such as coal, which, though now of comparatively little importance, may become more important in world trade in the future. Finally, adequacy can also be evaluated in terms of the capacity to serve particular destinations; adequacy presupposes some flexibility to deal with possible unforeseen developments requiring logistical changes or rearrangements.

Central to the questions of ownership and control are the avowed intention of the petroleum-exporting countries to enter the transportation phase of the oil industry, and the consequence of such a change should it actually occur. The magnitude of producer participation will be important; and the sectors in which they elect to concentrate their activities will be important.

A second consideration deriving from the possible shift of ownership of the tanker fleet to the oil-exporting countries has to do with the importance Western governments traditionally have placed on maintaining national shipping and shipbuilding capabilities. Because oil trade has such a commanding position in world maritime trade (49 percent in 1975), developments in oil transport have important implications for the viability of national shipping industries.

Beyond concern with the tank fleet per se, but intimately related to the logistics of supply, is the question of the security of the sea-lanes. Concern with the security of existing routes should be supplemented with an appraisal of alternative routes and the implications for (1) the defense of alternative routes, and (2) the speedy delivery of oil if alternatives must be used. In addition, the potential for U.S. interdiction of foreign supplies to third parties is of interest.

Finally, the security and defense of port and terminal facilities in both producing and consuming areas are of strategic importance. Port and terminal facilities could also be assessed in terms of their adequacy, i.e., their ability to process exports and imports in quantities sufficient to meet national requirements. The security of such major oil-exporting terminals as Saudi Arabia's Ras Tanura and Iran's Kharg Island is essential, as is the security of the major reception terminals in the oil-consuming countries.

Analysis will focus on these four areas:

1. adequacy of the tank ship fleet
2. ownership and control of the fleet
3. security of the sea-lanes
4. adequacy and security of port and terminal facilities

Adequacy of the Fleet

The current tanker fleet has very large excess capacity. In

the depressed market conditions of 1975, surplus capacity burgeoned to 114 million deadweight tons (DWT)—or about 40 percent of available capacity. In the first quarter of 1977, the surplus is estimated to be 100 million deadweight tons (DWT)—a commercial disaster reflecting a loss in demand from the worldwide recession and gross overbuilding.

In mid-1976, the world tank ship fleet totaled 306 million DWT. In spite of cancellations of orders for a number of new ships and the scrapping of some 13 million DWT in 1975 and the first half of 1976 (compared with 10.5 million DWT during the previous ten years), the tanker surplus remains serious. In addition, in mid-1976, the total order book comprised 56 million DWT. With no additional scrapping, the 1977 world tank ship fleet could equal 362 million DWT.

The adequacy of the fleet can be assessed only in relation to the quantities and types of commodities that the fleet will be called upon to carry and the destinations it will be called upon to serve. But the seriousness of the surplus capacity situation can be demonstrated by an example.

If Free World oil demand of 61-68 mmb/d in 1985 is assumed and if 60 percent of Free World consumption flows in world trade, the tanker fleet would have to move between 43 and 48 mmb/d, requiring a capacity of 172-192 million DWT. Under the worst possible conditions (which are extremely unlikely), i.e., no new construction beyond 1977 and the retirement of all tonnage built prior to 1971 (which would be 15 years old in 1985), the tanker fleet would still total 231 million DWT in 1985, more than adequate to meet world oil trade needs.

In the longer term, the surplus may narrow. On the demand side, forecasts may well prove to be too low, and oil demand may increase with stronger economic performance in the industrialized countries. Deepening U.S. involvement in the long-haul market will also result in greater tanker demand. However, only the smaller vessels (less than 80,000 DWT) can service existing U.S. ports, and while greater tanker supply-demand balance may develop for vessels of this size (some analysts forecast shortages in tankers of this size as existing tonnage gets old and is scrapped and as new construction is limited), the surplus in Very Large Crude Carriers will continue to plague the industry.

On the supply side, new construction will continue but at a reduced rate; high replacement costs (combined with low freight rates) do not encourage the building of new ships. High replacement costs also discourage scrappings, but the pressure for accelerated scrapping is inherent in the surplus capacity situation. Oil demand higher than anticipated, the beginning of exploitation of new reserves, reduced new construction, and increased scrapping could reduce the surplus.

But several factors could reduce this tanker surplus: (1) the increase in short-haul oil (North Sea and Alaska); (2) the possibility that oil-consuming nations will encourage the development of domestic or more proximate oil resources; (3) the possibility that nations, particularly the United States will enact national shipping preference laws necessitating the building of new tonnage by national shipbuilders; and (4) the intention of the oil producers to enter the transport business (to the extent that this involves new construction rather than the purchase or chartering of the existing surplus). In addition, the continued inability of the independent Tanker Owners' Association and the International Maritime Industry Forum to reach agreements on remedial actions does not augur well for reducing the surplus or "stabilizing" the tanker market.

On balance, it appears likely that that trend toward reducing the surplus will continue, but it will be slow and the oversupply phase will prove persistent. After 1985, if a tanker shortage were to develop, increasing freight rates would exacerbate the tighter oil supply situation anticipated at that time. The important point, in the short-to-medium term, is that constraints on crude oil supply will not originate in a shortage of tanker capacity. Instead, excess capacity, while narrowing, will persist.

The situation for the United States, however, with its dependence on the smaller tankers, which are old and are not being replaced, is different. The United States lacks the superport facilities now serving Western Europe and Japan. The only deepwater U.S. port capable of handling 150,000 DWT tankers is Long Beach, California; Cherry Point, Seattle, and Los Angeles are limited to 125,000 DWT. Not until the early 1980s, at present planning levels and commitments, will the United States be considered able to take economic advantage of the VLCC. The LOOP and SEADOCK projects would then give the United States two Gulf facilities capable of receiving 500,000 DWT tankers. Until then, its dependence upon the smaller and medium-size ships below 80,000 DWT will be extreme.

The situation with regard to product tankers is also different. It has already been suggested that an increasing share of internationally traded oil will take the form of oil products rather than crude oil. A tight refining situation on the U.S. East Coast and the continued expansion of refining capacity in the oil-producing countries will result in increased demand for specialized product tankers. One source suggests that the demand for products tanker capacity will reach 43-73 million DWT by 1980 and may reach 90 million DWT by 1985. According to this forecast, severe imbalances may develop in the late 1970s or early 1980s. In 1985, the overall shortage in products tankers may reach 63 million DWT, which will be diminished only to some extent through conversions.

It has also been suggested that dependence on imports of petroleum products is potentially more serious than dependence on imported crude when refining capacity is at normal near-full utilization and that it is difficult to shift rapidly from one refining source to another. To the extent that oil exporters expand their capacity in the transport area by increasing their products tanker capability, the degree of producer control over products may be even more serious, i.e., even if spare refining capacity is available, there may be no logistical element to take advantage of it.

Natural gas remains very much a "local" energy source. In the absence of currently available logistics systems (pipelines, LNG tankers, and processing facilities), it is tied closely to the location of production. Given declining U.S., Canadian, and, probably, Dutch production in the next decade and given a demand for large volumes of natural gas, more natural gas will find its way into international trade. Much of it, however, is expected to move via pipeline from Iran and the Soviet Union to Western Europe.

At the end of 1975, there were twenty-eight LNG carriers in service (capacity 1.7 million cubic meters). New LNG vessels on order total thirty-nine, with an aggregate carrying capacity of over 4.9 million cubic meters. However, inadequate LNG tanker capacity is not the only factor limiting LNG trade. Serious technological problems inhibit the growth of LNG trade, as the Algerian experience warns, and there are staggering increases in capital costs.

A final word on the adequacy of the world tanker fleet has to do with the continuing trend toward larger tankers. In mid-1976, 67 percent of the world tanker fleet consisted of ships of 100,000 DWT and over, and 82 percent of ships on order are 100,000 DWT and above. Indeed, ships of 200,000 DWT and above constituted 53 percent of the tanker fleet and approximately 64 percent of the ships on order in mid-1976. The trend toward larger ships means that economies of scale may be realized; it also may mean less flexibility if rearrangements in the logistics of oil supply are needed. Increased quantities of oil can be delivered to a single location, but the number of locations serviced declines.

In conclusion, adequate general crude oil tanker capacity seems assured for some time to come. In the area of products trade, the short-term outlook is not bright, but building programs could remedy the situation. Tanker availability alone, however, is not a sufficient condition for expanding world LNG trade where supply constraints arise from technological and economic factors. Finally, because the tanker size-mix may have some bearing on the flexibility and adaptability of the fleet, the future size distribution of the fleet will be of considerable importance.

Ownership and Control

Excess capacity has driven freight rates down. The de-

Table 9. World Tanker Fleet at End 1975
(vessels 10,000 dwt and over, million dwt)[1]

Flag	Oil company	Private	Government and other	Total	Share of total (percent)
Liberia	26	64	0.3	90	31
Norway		25	.2	25	9
United Kingdom	22	11	.2	33	11
Japan	4	27		31	11
United States	4	5	1.0	11	4
Panama	5	4		9	3
France	9	4	.1	13	4
Greece		16		16	6
Other Western Europe	14	22	.3	36	12
Other Western Hemisphere	6		.2	6	2
U.S.S.R., Eastern Europe, and China			8.0	8	3
Other Eastern Hemisphere	5	7	.2	13	4
Total	95	185	10.5	291	100

[1] Excluding 43,600,000 dwt of combined carriers.

Source: British Petroleum, Statistical Review of the World Oil Industry, 1975.

pressed tanker market results in two conflicting trends. On the one hand, tankers are available for sale at relatively low prices. On the other hand, the tanker business hardly looks inviting. Earlier, but now also, in the midst of these contradictory pressures, the oil producers have repeatedly indicated their intention to assume a role in the transportation end of the international oil industry.

Data on the "effective" nationality of the world tanker fleet are extremely difficult to obtain. Instead, tankers are often registered in certain foreign countries that specialize in providing favorable tax treatment for this type of business activity. Much of the world tanker fleet sails under "flags of convenience," particularly Liberian, Greek, and Panamanian flags.

It is possible, however, to identify the U.S.-owned and allegedly "controlled" fleet sailing under both the U.S. flag and flags of convenience.

The U.S. share is over one-quarter of the world tank ship fleet.

The Soviet fleet is some 3 percent of the world total and has the capacity to transport 2 mmb/d of oil. Given Soviet self-sufficiency in oil and the use of pipelines to serve Soviet customers in Eastern and Western Europe, the Soviet fleet is

Table 10. U.S.-Owned Tanker Fleet 1974 (million dwt)

Flag	Number	Million dwt
United States	306	10
United Kingdom	84	9
Panama	116	5
Liberia	411	36
All other	135	9
Total	1, 052	69

probably adequate to meet current needs, (including the supplying of Cuba). The Soviets are expanding their tanker fleet,
perhaps in anticipation of domestic requirements for Middle
East oil. However, Soviet intentions are unclear, since the
DWT of the ships envisioned in the Soviet plan (300,000
DWT) exceed the capacity of Soviet Black Sea and Baltic
ports. Thus, the Soviets may enter the world tanker market
for shipments between two foreign ports. It is not anticipated
that Comecon tonnage seeking employment in the international tanker market will be large enough to affect the market
structure.

Instead, the Soviets are probably preparing for two things.
First, as has already been demonstrated, some increase in
Soviet imports from the Middle East is likely, and enlarge-
ment of port facilities will occur. Second, Soviet spare capacity could be used strategically to support friendly producing
or consuming governments in conflicts with the Western oil
companies, other tank fleet owners, or particular consuming
governments. The existence of the Soviet option may at some
time be important (as it has been in the case of Cuba).

While the Comecon fleet is firmly under government control, the same cannot be said of the Western fleet. In 1976,
33 percent of the world tanker fleet was owned directly by
the oil companies. It is also likely that these same companies
dominated the private charter market as well (64 percent of
the fleet). Only about 4 percent of the world tanker fleet was
owned by governments; excluding the communist countries,
less than 1 percent of the total Free World tanker fleet was
government-owned.

A significant element in the continued role of the private oil companies in the producing states relates precisely to their ability to manage the complex logistical supply system in an efficient manner. In the long term, this ability may be a wasting asset, but since development of an OPEC or OAPEC tanker capacity will probably be slow, this oil company role may be significant for some years to come. In fact, the oil producers may prefer to refrain from potential competition among themselves and from the need to make hard allocative decisions by permitting the oil companies to continue to perform this function.

As more and more oil is turned over the the national oil companies for direct sales, the oil producers will both expand their own tank ship fleets and enter the tanker charter market. Still, in spite of their declared intention to expand their participation in the world tanker fleet, the oil-producing states apparently have taken relatively little advantage of depressed prices for existing tankers to enlarge their fleets. In line with what appears to be a partiality for safe investments, the producers remain cautious. Current forecasts suggest a rapid expansion of producer-owned tanker capacity, but their involvement continues to be only a small fraction of the world fleet.

The total OPEC fleet represents 3 percent of the total 1975 world tank ship fleet. The OAPEC fleet is 88 percent of the OPEC fleet. The OAPEC-sponsored Arab Maritime Petroleum Transport Company (AMPTC), founded in 1973 with an authorized capital of $500 million, recently placed orders worth $240 million for the construction of four supertankers to form the nucleus of the AMPTC Fleet. Even so, an anticipated Arab fleet of 20 million DWT by the early 1980s will represent only a marginal contribution to the world tanker fleet.

However, entry into the charter market is certainly a possibility, and specialization in particular sectors of transport, i.e., products carriers and LNG vessels, would increase the importance of the producers' fleet (if in fact they move toward specialization—a trend not yet certain but anticipated).

Table 11. OPEC Countries: Tanker Fleets, End 1975 (millions of dwt)

Country	Existing	On order	Total [1]
Kuwait	1.3	1.2	2.5
Saudi Arabia	.7	.3	1.0
Iraq	.4	1.5	1.9
Abu Dhabi	.5		.5
Unspecified Arab [2]		.8	.8
Iran	.7		.7
Libya	.4	1.0	1.4
Algeria	.3	.4	.7
Venezuela	.4		.4
OAPEC	3.6	5.2	8.8
Total	4.7	5.3	10.0

[1] Does not sum due to rounding.
[2] Country of registration not yet decided.

Source: John I. Jacobs and Co., Ltd., World Tanker Fleet Review.

Algeria is concentrating on LNG carriers; by 1979, it will own 10 percent of the world LNG fleet capacity. Saudi Arabia has indicated its interest in products tankers. Concentration on products tankers would also enhance the position of the Arab fleet, particularly since shortages are forecast in this area.

While the total capacity of the producing countries' fleet is unlikely to make more than a marginal contribution to the total capacity of the world fleet, concentration in particular areas could give this fleet economic and political importance. Producer governments could relieve themselves of the burden of world surplus capacity by requiring use of their flag/charter tankers. Preference for such tankers is an avowed objective of producer governments, even though it has not yet been applied in a manner that seriously affects either the economics of supply or its security.

Control over transportation has traditionally been considered a vital link in the chain of integrated oil operations. Interestingly, in terms of producer government participation in the world tanker fleet, it seems certain that expanded investments in oil transport would raise the cost to the producers of any future embargo or purposeful supply disruption. To the extent that the oil-exporting countries expand

Table 12. Employment of Tankers, 1975
(percent of world's active oceangoing fleet on main voyage)

Voyages to—	Voyages from—					
	United States	Caribbean	Middle East	North Africa	Others	Total
United States	3	3.0	6.0	1.0	3.5	16.5
Canada		.5	3.0			3.5
Other Western Hemisphere			6.5	.5	2.0	9.0
Western Europe, North and West Africa		1.0	42.0	1.5	3.5	48.0
East and South Africa and South Asia			1.5			1.5
Japan			11.5	.5	2.5	14.5
Other Eastern Hemisphere		.5	4.5		.5	5.5
U.S.S.R., Eastern Europe, and China			1.5			1.5
Total	3	5.0	76.5	3.5	12.0	100.0

Source: British Petroleum, Statistical Review of the World Oil Industry, 1975.

their role in the oil supply logistical system, they may increasingly share the consumers' interest in the smooth and continuous flow of oil to market.

National Shipping Industries

Depressed oil demand, excess capacity, low freight rates, and the increased participation of the the oil producers in the world tanker fleet have raised serious questions about the viability of national shipping and shipbuilding industries. As long as the surplus capacity situation persists, national shipping industries will be under additional strain. Historically, governments have considered a healthy shipping industry to be a necessary element of national security policy. It is difficult to think that this attitude will change. Subsidies and national preference laws may proliferate, and the potential for international conflicts among these laws will increase over time.

Security of the Sea-Lanes

The importance of the sea-lanes from the Middle East to

Figure 2. Principal International Oil Movements.
Source: British Petroleum, *Statistical Review of the
World Oil Industry,* London, 1975, p. 15.

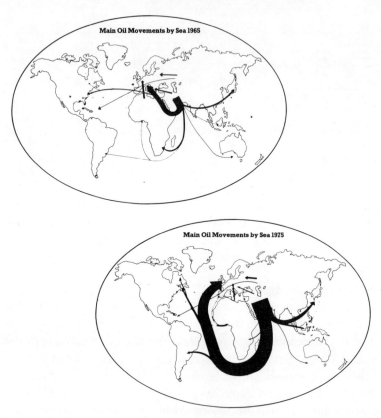

the United States, Western Europe, and Japan is obvious.
Over 75 percent of the world's active oceangoing tanker fleet
is engaged in transporting oil from the Middle East to the rest
of the world. Fully 66 percent of the world tanker fleet is
engaged in the transport of oil from the Middle East and
North Africa to markets in the industrialized world, i.e., the
United States, Canada, Western Europe, and Japan.

Given the oil supply-demand outlook, the importance of

Gulf oil will increase. Direct U.S. interest in the Gulf will increase as the United States imports more oil from this source. Even in the unlikely event that U.S. imports from the area do not increase, Gulf oil will remain of vital interest to the United States; U.S. allies in Europe and Japan will continue to be heavily dependent on this source.

Within the region, the Straits of Hormuz are of the utmost importance. The overwhelming share of Iranian, Kuwaiti, and Saudi oil must pass through the Straits on its way to market. Moreover, the Straits are vital because existing (and probably future) pipeline capacity for moving oil through the Mediterranean terminals is inadequate to the task of transporting the huge quantities involved. For much Gulf oil, there is no alternative to shipment through the Straits of Hormuz to markets in Western Europe, Japan, and the United States.

On the face of it, there is little to inspire much confidence in the continued stability of the Gulf. The continued viability of the small Arab sheikhdoms united in the Union of Arab Emirates remains to be demonstrated. Iraqi relations with most of its neighbors are poor. Iraqi-Saudi hostilities are barely concealed; Iraqi-Kuwaiti relations can be nothing but poor given Iraqi irrendentist ambitions; and it may be too soon to declare Iraqi-Iranian hostilities a thing of the past. There were also indications that the Iraqis and the Syrians confronted each other on opposite sides in the Lebanese civil war.

In addition, Saudi-Iranian relations are complex. The two nations share common interests in opposing radical regimes, upholding the monarchial tradition, and protecting the oil flow. However, the (Arab) Saudis are not enthusiastic supporters of the (non-Arab) Iranians, and the Saudis are clearly reluctant to see a growing, unopposed Iranian military capability in the area. The situation is further complicated by the fact that both nations are U.S. clients. With regard to the Gulf (which the Iranians call the "Persian Gulf" and the Saudis term the "Arabian Gulf"), the coincidence of certain interests should not obscure the real and basic lack of trust between the two countries.

In the event of armed hostilities, not all of these simmering conflicts imply a complete closure of the Straits. If the Saudis had the capacity to send increased amounts of crude out through Tapline and through a pipeline to be built to a new Red Sea port, an Iranian-Saudi conflict might see the Saudis trying to starve Iran by disrupting oil flows through the Straits. Even in the absence of such alternative Saudi export routes, closure of the Straits would be much more serious for Iran than for Saudi Arabia. At present the Saudis do not have the military capacity to oppose Iran in this way, while the Iranians could oppose the Saudis by barring Saudi traffic through the Gulf, perhaps in an effort to prevent possible increased Saudi oil exports from exerting downward pressure on prices.

Indeed, Iraqi-Kuwaiti, Iraqi-Saudi, or Iraqi-Iranian armed hostilities might leave the oil flow through the Straits unimpaired. Radical regimes in the sheikhdoms might allow terrorist and sabotage activities from their territories, but this would surely open the way for swift Saudi or Iranian military intervention.

At second glance, then, the Gulf and the Straits do not appear as insecure as a first reading might suggest. Currently the states in the region reject a greater direct American military presence in the area: it might provoke the very type of external interference they wish to avoid. Iran can probably defend the Straits against any equivalent power in the area not given large external assistance. But it is not likely the danger would come from such a source. The United States must therefore make the American security commitment to the Gulf so exceedingly continuous, clear, and firm as to avoid any possible miscalculation by the Soviets and the Chinese (who apparently were active in the Dhofar rebellion in Oman). The U.S. defense objectives with regard to oil could have more to do with European and Japanese dependence than with that of the United States itself; they are nevertheless of crucial importance. A related energy objective is, of course, to preclude a Soviet advance upon the oil reserves.

If the Suez Canal were used to transport additional quantities of crude, tankers would not necessarily avoid the Straits of Hormuz. The Canal does not appear to be an attractive route today—freight rates are down, and the Canal cannot handle larger ships—but it should be recalled that the Suez Canal once handled virtually all Middle East oil shipments to European destinations. Present Canal deepening and widening anticipates 53 feet depth by 1978 (150,000 DWT laden, or 270,000 DWT tankers in ballast) and 67 feet depth for 270,000 DWT tankers fully laden by 1982. If necessary, the Canal might be called upon again, particularly since it would shorten the supply line to Western Europe. While the Canal is undoubtedly far less important today than it once was, the security and defense of the Canal deserve continuing attention. This in turn means a U.S. defense commitment to the Red Sea and the Mediterranean as well.

Recent developments in the Red Sea area are far from encouraging. Ethiopia, firmly tied to the United States before the 1974 coup, is the scene of continuing internal political disarray. Tribal, ideological, and personal power rivalries continue to plague the country, and there is warfare in the streets between the Marxist military regime and its various opponents. Along the northeastern Red Sea coast, Eritrean rebels continue to seek separation from Ethiopia. In the northwest, remnants of the army loyal to the old regime continue to oppose the new military regime. In the south, Somalia continues to harbor irrendentist ambitions concerning Ethiopian territory, and the two countries are on a collision course over the French territory of Afors and Issas, which was scheduled for independence in June 1977. The military government has also antagonized Sudan by providing opponents of the current regime with a staging base. Finally, the Ethiopian government has apparently welcomed Soviet interest in the country.

The Soviets, allied with Somalia and maintaining a military base in that country at Berbera Ras, recently sought to reconcile the two countries. President Castro, on a recent African trip, allegedly attempted to gain Ethiopian and Somalian support for a federation that would eventually include those two countries plus South Yemen and the territory of Afors and Issas. In this way the Soviets hope to build stronger ties to Ethiopia without jeopardizing their position in Somalia, where they have what may be their largest investment in Africa, especially in military supply and base facilities. If they succeed, they would surround the entrance to the Red Sea with countries aligned with the Soviet Union, and this would put potentially hostile countries on the flank of tanker lanes all the way from the Gulf through the Indian Ocean.

In opposition to the Soviet scheme, Sudan, with the backing of Saudi Arabia and Egypt, has attempted to woo Somalia and South Yemen, with promises of substantial financial aid, away from the Soviets and toward the moderate Arab camp. In addition, there may be Arab support for the Eritrean nationalist movement. Should the Eritreans succeed in their secession attempt, they will take with them the entire Red Sea coast of Ethiopia.

The stakes are high—essential control of the Red Sea—and at this writing there is no resolution in sight.

If the United States itself takes increasing quantities of Middle East and African oil, the very long supply lines around Africa and across the Atlantic must be kept open. The sea-lanes between Venezuela and the Caribbean refineries and the U.S. mainland must be maintained, a task possibly of great difficulty, as World War II demonstrated. The Panama Canal is of little significance today for the oil trade, but this might not always be true—especially if Alaskan crude moves by sea to the U.S. Gulf or if, in time of war, the supply route from Indonesia to the U.S. West Coast were severed (this would require westbound oil traffic).

For U.S. supply, then, assuming a diminished dependence on the Persian Gulf, the area of special concern would be the supply lines from Alaska, the Caribbean, and Nigeria. These would be almost irrelevant to European and Japanese concerns—a reminder that for U.S. relations with its NATO allies and Japan, our respective, varying degrees of dependence upon Persian Gulf crude are very significant.

For Japan, the Straits of Malacca, which separate Malaysia and Indonesia, must be kept open. The Indian Ocean and the China Seas are also important. With regard to the Indian Ocean, perhaps as much as 85 percent of Japan's oil requirements must cross this body of water. The growing Soviet presence in the area over the last few years is thus of concern. Compared to the normal complement of seventeen Soviet warships in the area (sometimes twice that number are there), the United States has a permanent force of only three ships in the Indian Ocean, stationed at Bahrain under an agreement now being renegotiated. The United States is in the process of building a naval base on the island of Diego Garcia. The Soviets and the United States continue to hold discussions regarding the control of the arms race in the Indian Ocean.

The growth of Soviet naval power is obviously a matter for deepening concern. Over the past decade, the USSR has increased its presence in the "choke points" close to the Gulf. Its naval visits, facilities, and "client states," real or potential, can permit its units to locate at virtually any segment of a tanker voyage from Hormuz south through the Mozambique Channel. An overt action by Soviet naval units to sink tankers would have such extreme consequences that it would probably be made only as a prelude to general war.

A more likely security problem would come from terrorist or revolutionary political groups, who with comparative ease could attack a tanker from so many places as to make protective measures for individual tankers almost impossible. Small vessels with conventional weapons could attack tankers in ballast and possibly destroy them without warning or without any message from the ship. Several VLCCs have disappeared in recent years—as a result, it is thought, of internal explosions. If others were to disappear, the consequences would be far greater than the loss of a single ship. Insurance rates would undoubtedly rise to wartime levels; there would be an immediate slowdown in tanker schedules and diversions. In the case of Japan, for example, or in the case of any heavily import-dependent nation, a delay of a week's scheduled deliveries—in the absence of readily available crude stockpiles—could cause immediate shortages; some 30 million barrels would not have arrived. Action by terrorist revolutionary political groups could well have an unprecedented effect.

U.S. interdiction possibilities, so far as the Soviet bloc is the target, are currently limited on the seas. As matters stand today, the Soviet Union is self-sufficient in oil, and Soviet oil exports are for the most part transported by pipeline to Eastern Europe. In the future, tanker transport may be of increasing importance as the Soviet bloc finds itself in the Middle East oil market. Imports are likely to originate in Iraq for tanker shipment to Eastern Europe. Some Libyan and Syrian oil may also find its way into the Soviet-bloc market.

Security of Terminal Facilities

Finally, brief mention should be made of the unprecedented vulnerability of the United States, in particular, to the loss of a superport facility—such as LOOP or SEADOCK. This could make it extremely difficult and perhaps impossible to receive a very large share of overseas oil. Unless there were many more VLCC receiving terminals on all coasts, the loss of two giant discharge points could conceivably be as serious as a massive embargo launched by key producers. While defenders of LOOP and SEADOCK argue that their loss would be an act of war, the real point is that the loss of LOOP or SEADOCK would be extremely serious in the event of war or otherwise, as increasing volumes of U.S. oil imports are channeled through these facilities.

"New" Oil

The continuing importance of oil in the economics of industrial nations is, as indicated earlier, assured. Neither efforts to increase coal production for domestic consumption or for world energy trade nor efforts to produce synthetic fuels or natural gas will significantly affect the dominance of oil. Nuclear energy will have such an effect but probably not until close to the end of the century. For the next quarter-century, oil will remain the single most important source of energy.

While the search for new oil reserves in other areas will intensify, the Gulf region will remain the single most important source of oil entering world trade. Since the area is likely to be subject to acute pressures from states external to the region and to issues between Gulf states, its "stability" will be an ever-present question of widespread concern.

Unfortunately, stress is still placed—mainly in the United States—on the need for new, but "non-OPEC," sources. The hunt will be unproductive. Whether a particular discovery is found in an OPEC state, or whether, once the discovery is made, the country is accepted into OPEC, or whether the country takes advantage of OPEC pricing—all this may be quite irrelevant. The example set by OPEC will be quite enough so far as pricing is concerned. Moreover, the danger of a sustained "OPEC" embargo is considered remote. OAPEC is a more likely source of action. Therefore, the search will be for oil outside the Gulf.

The need for such a search is unarguable. Serious shortages, which are expected to appear about 1985 and which may be of mounting severity, will be reflected in the falling reserve-production ratio discussed earlier in this report. To meet "minimum" requirements, several sources warn, the world's proved reserves of oil should total at least 800 billion barrels by 1985; that is 225 billion barrels more than was on hand in 1970.

In order to avoid dependence upon the Middle East generally and thereby to reduce vulnerability to interruptions, proved reserves in other areas will have to be more than doubled. The need, then, is for 600 billion barrels of new oil by 1985—or nearly 50 percent more oil than has been found and developed since 1955. Of the 445 billion barrels actually found and developed since 1955, only 106 were discovered

outside the Middle East and Africa. (Note, also, that anticipated needs met, and to be met, in the period 1970-1985 will
have consumed fully two-thirds of all the proved reserves in
existence in 1970.) Does such "new oil" actually exist? No
one knows.

Most petroleum (outside the Middle East fields) has been
discovered in areas once thought to be barren or beyond
reach. Knowledge of the geology of petroleum continues to
expand. The requisite technology to extract oil grows expo-
nentially. Many factors can explain the failure to improve
upon the rate of discovery: inadequate knowledge of geology, insufficient testings, poor appraisals, lack of corporate
interest in certain regions, costs, inhospitable political conditions, and plain bad luck. Each factor will continue to plague
exploration and exploitation.

At some juncture, we may conclude there is in all probability no oil recoverable in sufficient volumes. Perhaps that
signal is already being flown, perhaps we may see the signal
years from now, perhaps we may "never" see it. All that one
can prudently note is:

1. The need for enormous new reserves is now with us and
   will grow as each year passes without the requisite rate
   of additions to reserves. In considering 1970-1985, six
   years have already passed without very large finds outside the Middle East. We have nine to go until 1985.
2. Competition for available supply will become intensified
   over the coming years, even if huge new discoveries are
   made: the oil cannot immediately be brought to market,
   nor of course can a field be brought quickly to maximum production level and kept there.

In attempting to guess the location of significant resources,
extreme caution is the rule. The actual amount of data available to substantiate a finding as to where prolific resources
may exist can range from nearly zero to reasonably complete
banks of knowledge; unfortunately, by far the greater part of
offshore areas lies in the "nearly zero" category. Ultimately,

it is still necessary to poke a hole, and then another. . . . As only one example of the great range in estimates—and it is important, for it has to do with offshore undiscovered oil resources—we cite two essentially conservative guesses: one indicates perhaps 50 billion barrels, the other, equally authoritative, indicates 56-120 billion barrels. We are only beginning to find out which may be closer to the "truth." We will not know until the end of the century. Nevertheless, because these resources lie close to U.S. shores, their exploration and exploitation is generally acknowledged to be of very high priority. (If anything is needed to underscore the importance of a successful conclusion to the resources sovereignty/jurisdiction issues in the Law of the Sea negotiations, it is these offshore and basin areas.)

Moreover, current forecasts deal very largely only with offshore regions, reflecting the widespread belief, which amounts to a conviction, that great prospects lie under the oceans, not under land. Whether this is "true" or whether the petroleum industry is acting like sheep (as it usually does in these matters) cannot be known. What is worth remembering is that not one of these forecasts would have been given such attention fifteen years ago.

In 1975, offshore production was a total 357 million tons (6.8 mmb/d), or 14 percent of the world's total production. Surprisingly, 1973 and 1974 showed considerably larger offshore production: 503 million tons and 463 million tons, respectively. The unexpected decline may be due to deliberate commercial policies to produce more from onshore fields, higher offshore costs, lower demand, as other factors; we cannot tell. Nearly all current offshore production comes from fields that are extensions of shore deposits. The following are widely regarded as of prime prospective interest: the Arctic, the north and east coasts of South America, the rim of the Caribbean, the area offshore the northeast United States; northwest and west Africa, both shores of the Mozambique Channel, the Bay of Bengal; the west coast of Malaya, the Surabaya Sea, the west and south coasts of Australia, and east China.

It is beginning to be thought that very large deposits may be found in the comparatively small marginal basins, which received the sediments of great rivers; these are not part of shore deposits. Moreover, recent geological searches have resulted in wholly tentative observations that staggeringly large sedimentary deposits lie on the continental rises. The size of these deposits may be as much as half of all sediments deposited anywhere. Even abyssal plains and mid-ocean regions such as along the Atlantic Ridge show some such deposits.

None of these has been subjected to exhaustive surveys, let alone the essential test of drilling; none is presently being surveyed comprehensively, and it may be that few will be—given the vast expense and technological advances required. It is clearly a task that government may have to help underwrite and, if necessary, undertake; the need to tap very large reserves is, as we have stressed, of greatest urgency. The gambles involved may lie beyond the financial resources of the industry.

One of the most promising basins is in the Gulf of Mexico. "Confirmed" reserves of 60 billion barrels are not unlikely; these reserves could sustain a 2.75 mmb/d production rate in 1980 and possibly 6 mmb/d in 1985. Other areas needing the earliest possible intensive study and exploration lie offshore Alaska (and the National Petroleum Reserve).

The northwestern and northeastern Canadian Arctic may not be of highest priority, but parts of offshore Greenland and the upper areas of the North Sea into the Spitsbergen region could warrant whatever costs it may take to survey and explore. Middle-range forecasts of exportable surplus from Norway and the United Kingdom's combined share of North Sea oil could amount to 4.5 mmb/d by 1985, 6 mmb/d by 1990, and even more.

All of these sites may not be as well endowed as the Soviet Arctic from the Barents Sea to the Bering Strait. But they are either under direct U.S. or industrial nations' control, or else have comparatively short logistic lines. A comparable effort with Venezuela is necessary to determine the feasibility of exploiting to our maximum mutual benefit the tar belt of the

Orinoco. Similarly, despite the evident political difficulties presently before us, an energy effort with Canada would seem to be of obvious priority.

Even moderate success in tapping these hemispheric sources would alter the geopolitics of oil for the United States and hence for the other industrialized countries as well. Given the long lead times involved in the exploration, development, and exploitation of these areas, they are possibilities only in the long term.

3
Coal, 1977-2000

Coal is nearly always touted as the world's energy resource whose more rapid exploitation could offer the prospect of energy autarky with (or even instead of) increasing application of nuclear power. For example,

If energy consumption from all fuels were to grow at the end of the present century at the annual 5% rate . . . cumulative energy requirements to the end of the century . . . might amount to 400 billion tons of coal equivalent. Not only could estimated 4.3 trillion tons of estimated recoverable coal resources meet this entire growth of energy demand, but in the year 2000, at then prevailing rates of total energy consumption, enough coal would be left in the ground to meet the entire bill for a century and a half beyond.[1]

On a less dramatic note, coal is still considered to be the readily available alternative to oil imports. It is our view that coal can in no realistic assessment substitute for oil imports (nor for nuclear energy). However, substantial increases in coal production could importantly diminish the United States' continued dependence upon imported oil. For that reason alone, coal needs to be exploited.

At the end of the century, the increasing use of nuclear power—not coal—may be the key factor displacing oil in electric power generation, the most critical growth sector in U.S. national energy consumption.

Coal Resources: Location

The figures for coal are awesome.[2] It is largely found in the northern latitudes, but its present and prospective importance in world energy trade is minimal.

Nearly 90 percent of the world's total resources of coal exist in the USSR, the United States, and China, most of it above 30° latitude. It is, preeminently, the energy resource under the direct control of the industrial world; the amount possessed by LDCs is negligible (however valuable it may still be as an indigenous energy resource for a particular LDC). The volume of coal in world energy trade is limited largely to U.S. exports (presently some 10 percent of U.S. domestic production). As a percent of energy (BTU) in world trade, coal is generally unimportant; Japan and Canada are overwhelmingly the major importers, with some 20 million tons and 17 million tons of imports, respectively.

The holdings of the three coal giants dwarf those found anywhere else in the world, and it is extremely unlikely that discoveries elsewhere will alter the relative importance of present producers for as many decades ahead as one can hazard a guess. For any of the great possessors of vast coal deposits—the Soviet Union, United States, and China—a 50 percent recovery rate would theoretically make available an amount of energy enormously greater than the total available from their oil and gas. Current total resource estimates for the world's coal are measured at some 10.8 trillion tons. Of this amount, some 1.4 trillion tons are considered to be known reserves, and 0.6 trillion tons are economically recoverable under contemporary prices and technological methods.[3]

The USSR itself probably has 50 percent of the world's total coal resources, the United States some 19 percent and China possibly has about 8 percent.[4] Substantial resources exist in Europe, South Africa, India, Indonesia, Australia, and elsewhere. Although these resources are not in any way comparable to those of the "giants," they do constitute a largely unexploited source of energy. A summary of the key coal de-

Table 13. World Coal Deposits (in megatons)

Country or region	Reserves in place [1]	Economically recoverable
1. U.S.S.R.	273,200	136,600
2. United States	363,562	181,781
3. Europe	319,807	126,775
4. China	300,000	80,000
5. "Oceania" (chiefly Australia)	74,699	24,518
6. Africa	30,291	15,628
7. Latin America	9,201	2,803

[1] Note that reserves in place rank the United States and Europe above the U.S.S.R. while the more general measure used earlier of total resources gives a clear lead to the U.S.S.R. While confusing, the reserves in place figures reflect greater knowledge of deposits and their potential exploitability than do the total resource estimates. In the fullness of time, we expect the U.S.S.R. will rank highest in all 3 categories: total resources, reserves in place, and economically recoverable.

posits clearly illustrates the overwhelming importance of certain nations or regions.

Obviously, such figures indicate orders of magnitude only. Although it is thought that data on coal are more reliable than for any other energy source, the *Survey of Energy Resources* explains that national formulas differ and, in each case, reflect judgments as to changes in technology, transportation costs, labor costs (and availability), government subsidies, environmental factors, population changes, and the economics of competitive fuels. (For the industrial world, for example, at least at the outset, the potential of the "breeder" reactor looms very large in considering coal investments, since the value of the "breeder" is seen presently to lie particularly in electric power generation.)

Coal Resources: Increased Supply

Almost every industrial state is now seeking to redress the declining share of coal in its energy balance; only fifteen years ago, however, the emphasis was on greater use of oil as a less expensive, far less labor-intensive, and generally much more convenient and cleaner source of fuel.

In 1960, coal represented 37 percent of OECD energy consumption; by 1965, it was 28 percent, and by 1975 it had de-

Table 14. Free World Supplies of Coal (in millions of tons)

	1975 (estimate)	Forecasts 1980	Forecasts 1985
United States	573	680	845
OECD Europe	320	309	297
Rest of OECD	120	160	180
Non-OECD	230	310	415
Total	1,243	1,459	1,737

clined to 19 percent. By 1980, assuming moderate success in achieving greater production and use of coal, its declining share in world energy consumption might be halted, and in 1990 coal could still represent about 20 percent of world energy consumption.

As with oil and gas, percentage declines in share of energy consumed do not imply decreased volumes produced; on the contrary, for as far ahead as one can see more and more coal will be produced. Some 400 million tons were mined in the United States in 1960; by 1990 nearly one billion tons could be extracted. However, it is reasonable from these figures and conservative forecasts to believe that near–energy autarky for the leading industrial states is not to be found from coal. Nor will increased coal production—conservatively estimated— eliminate dependence upon oil imports.

Such increases (with the exception of Europe), translated into barrels per day, indicate that the growth in coal supplies would by 1980 be the equivalent of about 3 mmb/d of oil and by 1990 represent another 3.6 mmb/d—for an estimated total additional supply of 6.6 mmb/d to be achieved over ten years. This increase would be no greater than the increase in Saudi production alone from 1971 to 1975. In other words, an increase in available energy on the order of 6.6 mmb/d in 1985 would be some 4 percent of the estimated QBTU consumption of the industrial Free World.

Increased U.S. coal production, translated into oil equivalents, would be about 3 mmb/d, or a possible one-third of anticipated 1985 oil imports; thus in the case of the United States (and probably it alone of all the leading industrial states of the Free World), increased coal production could diminish supply vulnerability. For OECD Europe as a whole, there is less prospect of such a role for coal: no increase in production is foreseen after 1975, the reserve base is smaller, and the cost is immense. Coal is found primarily in Germany and France, and each chooses rather to emphasize nuclear energy. Japanese dependence on imported oil cannot be significantly diminished as a consequence of increased world production of coal. It is not thought likely that increased production will enter world trade.

Since the increased production for the United States is all from domestic sources, it is clearly important that this increase be achieved—a forecast increase of 272 million tons by 1985, which is far below the current "conservative" FEA projected increase of 440 million tons by 1985.

The contribution of coal in the form of synthetic crude or gas, much heralded as a key factor in U.S. energy supply, has been downgraded continuously. Currently, it is estimated conservatively that U.S. production of these synthetics will not exceed 1 mmb/d by 1985 (about 2 percent of the national QBTU consumption). Twenty plants producing synthetic natural gas (SNG) at 250 mmcu feet per day could cost some $30 billion (1976) and are said to consume some 165 megatons of coal a year (in the process using up 4 QBTU). If so, they could then contribute only some 2 QBTU to the national energy balance in the form of gas (2 Tcf per year).

The FEA forecast has 1.06 Tcf from gasification plants (and an additional 1 Tcf of substitute gas from petroleum products, which are normally included in the general oil category and are thus not a net addition to energy supply).[5] On this scale, emphasis on synthetic natural gas from coal could be justified, possibly, only if it resulted in lower imports. If these forecasts are correct, the 2 QBTU contributed would be the equivalent of 1 mmb/d of oil, or some possible 10 percent of 1985 oil imports.

Despite its advantages for the conservation of conventional oil and national energy security, it will not be easy to increase U.S. coal production and to arrest the decline in Europe's. The problems in doing so are familiar, and the questions are of greatest importance to the timely revival of a declining industry. What of the comparative attractiveness of nuclear power (especially the "breeder")? Will coal be price competitive? Or will there be long-term government subsidization? What of the availability of labor? How to finance the improvement in coal extraction and processing? Who is to meet the urgent and basic need for a greatly improved and extended logistics system for coal? Will there be a modification of environmental standards? Will there be compulsory use of coal in industry and electric power generation? Nearly three years after the onset of the oil embargo, we are not close enough to any answers to any of the questions. In spite of the Carter administration's emphasis on further utilization of domestic coal resources, insufficient attention has been devoted to the fact that each of the considerations listed above are crucial to success; their timing and "interlocking" aspects make success in some, but not in others, insufficient.

4
Gas, 1977-2000

Natural gas, after oil and coal, is the third most important source of energy for the industrial world. Currently, significant production and the volumes in world energy trade lie wholly within the control of the industrial world; this is not always to be the case.

In the future, very considerable amounts of natural gas, enough to dwarf present producing sources, will come preeminently from the Middle East and the USSR; it is also possible that the Arctic may yet be one of the more prolific regions.[1] A major reason why there is not early anticipation of natural gas in world trade from the Middle East lies in the absence of an adequate logistics system—especially the absence of specialized tankers and onshore facilities to gather, reduce, and, at the other end, to receive and process gas for distribution. Were these available, natural gas could be an important source of energy for all the industrial world, but it would also compound the security implications of more energy dependence upon the Gulf states.

However, attractive as the commercial aspects of natural gas are, there is a critical security issue: natural gas production may be easily and simply shut down. Moreover, in contrast to oil, which has relative flexibility in sources, gas supply arrangements are usually considered to be limited to a particular source—for reasons of technical quality, the design of processing plants, and the immense capital sums that must be committed to the whole undertaking. In addition, because of the huge costs associated with LNG, there is a close meshing of supply to demand over many years; a producer seeks to have commitments for all that is available. If one source is shut down, it is not likely that another source could quickly substitute.

For the rest of this century, natural gas may not be very important in world trade. But for particular countries, or sections of a country, it can be very important indeed, so important as to render a sector of one's energy consumption peculiarly vulnerable to supply cutoffs. This could be true even when the volume of imported natural gas is a very small fraction of the total. For these reasons, a nation's dependence upon imported natural gas raises serious questions of energy security and must be under continuous review.

The Energy Resources Council policy statement on LNG imports cautioned that if all LNG applications pending before the U.S. government were approved (3 Tcf), plus the 0.4 Tcf already approved, U.S. regions importing supply could be 15-30 percent dependent on overseas sources for their gas needs. The Boston region, for example, could be over 40 percent dependent on LNG imports for its gas supply. In order to lessen the risk, the Energy Resources Council limited LNG imports from any country to no more than 1 Tcf, with a maximum of 2 Tcf a year imported from all sources; 2 Tcf would be about 10 percent of total natural gas consumption in the early 1980s. President Carter, acknowledging the security implications, has recommended that the 1 Tcf ceiling be lifted but that no further public funding be available for LNG imports—a curious and unexplained limitation.

The possibility that natural gas will be a major factor in world energy supply depends on as many factors as in the case of oil: decisions by producer and consumer nations, price levels (and "guarantees"), competitiveness, alternatives, adequate and timely investments in fields, transport and receiving terminals, technology, assurance of continuous supply, and others.

Location

Currently, while natural gas may be only some 6 percent

of the world's total recoverable fossil fuel resources, natural gas provides for nearly 19 percent of the world's production of energy. If the world experiences a 4 percent annual increase in demand for natural gas, it could take less than fifty years to deplete the resource to a level at which ten years of reserve might remain. Thus the natural gas reserve-production ratio is as important as it is in oil.

With declining gas production in the United States, a trend only partly and perhaps temporarily reversed by the Alaskan finds, and with the gradual depletion of the Groningen and North Sea fields expected to occur over the next several decades, it is not probable that the Free World's industrial states will meet their natural gas requirements out of their own or nearby resources. There are increasing prospects for growing reliance upon the still very largely untapped resources of the Middle East and the USSR.

Currently, the USSR may possess nearly 36 percent of the world's total gas reserves—it is easily the natural gas giant. The Middle East may have 24 percent, and between them the two have over half (Iran is usually thought to have nearly half of the Middle East's gas). The North American continent may have 13 percent, and West Europe may have 8 percent of the world's total natural gas. If these estimates are contrasted with natural gas consumption, the problem of a declining ratio between reserves and production outside of the Middle East and the USSR becomes clear: with 51 percent of the world's natural gas consumption, North America (overwhelmingly the United States) has 13 percent of the reserves; West Europe consumes 15 percent but has 8 percent of the world's total of natural gas; and the USSR consumes 22 percent but may have 36 percent of the total.[2]

Gas in International Trade

While Western Europe may now be considered "self-sufficient" in the sense of having enough natural gas to meet current demand, future provision of gas out of world trade holds large uncertainties. For example, since the United States has

less than eleven years' use at current rates of production, an
early Alaskan and possibly a Canadian Arctic contribution to
supply is of great importance to a nation whose energy
balance has gas at 30 percent. For Europe, Groningen alone
supplies 40 percent of total European supplies (natural gas is
15 percent of energy consumption now but may rise perhaps
to 19 percent by the early 1980s). In both cases—Europe and
the United States—whether the percent of energy consumption met by natural gas rises or falls, the volumetric demand
for gas increases. Natural gas represents about 18 percent of
world energy today; by 1990 it may be 13 percent, but its
volume could increase by 25 percent. Groningen could reach
its producing plateau by 1978; will gas production from the
southern North Sea decline in the next decade but be replaced by other North Sea production? Europe's growth rate
in the use of natural gas has been nearly 30 percent per annum; it seems as if Europe's future gas growth will inevitably
be constrained in the 1980s. Europe's natural gas "self-sufficiency" will no longer be 94 percent (1975) but perhaps
75 percent by 1985. For the United States, natural gas (and
SNG) might be only 17 Tcf—against 21 Tcf production in
1975—and with Alaska it might be near 19 Tcf in 1985. But
in 1985 the share of gas in U.S. energy consumption may be
down to 25 percent (30 percent in 1976).

Currently, the only significant gas exporters are the
Netherlands, Canada, Iran, and the USSR. In 1974 the Dutch
share of natural gas exports was 41 percent (to West Germany, Belgium, France, and Italy). Canada's share was 23
percent (all to the United States). The USSR exported 12
percent of natural gas in world trade: to East Europe and to
West Germany, Italy, Austria, and Finland. Iran's exports
were at 8 percent (to the USSR). The United States and West
Germany were the largest importers, each taking 25 percent
of natural gas exports. The USSR, Belgium, and France took
some 10 percent each. In 1974 LNG was 11 percent of world
gas trade, with Brunei supplying 40 percent of LNG, Libya
28 percent, Algeria 20 percent, and Alaska 11 percent. Japan
took over one half (all of Brunei and Alaska). These ac-

counted for nearly all of the world's international gas trade, an amount that was only some 10 percent of the world's marketed production. Thus international supplies of gas remain a supplemental and only a fractionally small piece of natural gas consumption; international sales were two-thirds of the natural gas flared. Unless some very extensive and relatively accessible new reserves are discovered or unless new transport technologies are developed, the future of gas may be limited and possibly confined to its natural "premium markets." It may be no substitute for oil or coal.

Nevertheless, the proved reserves of OPEC natural gas remain very significant—as potential sources:

	Tcf
Iran	200
Saudi Arabia	54
Iraq	20
Kuwait	42
Libya	27
U.A.E.	12
Algeria	106
Nigeria	40
Venezuela	36
Indonesia	6

For the United States, the probable maximum LNG imports until 1985 may have been already defined—2 Tcf (mainly from Algeria and a smaller amount from Indonesia)—and of this only 0.4 Tcf (Algeria) comes from a project already operational, with the others under construction. Inordinate delays from bad planning, poor technology and design, renegotiations, persistent uncertainties over price, U.S. regulatory delays, and the exceptional lead times involved raise doubts about the other projects. But again, depending on its geographic destination within the United States and on its specific end-use and local importance, even the lower fractional contribution (the 0.4 Tcf figure) could have security significance.

If all current projects now before the U.S. government (under construction and potential projects) were approved (Nigeria, Indonesia, Iran, and the USSR), their total natural gas import contribution would amount to 3 Tcf by 1985. By 1985 imports might be 2 Tcf for the EEC and 2 Tcf for Japan, assuming that all projects under construction are brought into production.

Estimates of future world LNG trade have been consistently reduced because of soaring capital costs and the diminishing discounts in the price of landed LNG vis-à-vis landed crude oil. A recent study by British Petroleum estimates that if the LNG export projects now under consideration were to materialize, the capital cost involved could approach $30 billion (1976 U.S. dollars), including the costs of liquefaction and regasification facilities and shipping, but excluding production and gathering costs.[3] If they were to materialize, furthermore, LNG in world trade by 1985, would be equivalent in calorific value to some 3 mmb/d of crude oil. These international contributions to energy balances will be dwarfed by oil in world trade.

The USSR

The prospect for Soviet natural gas could be very bright indeed. Not only may it be able tap huge quantities of natural gas for its own domestic requirements, but it should also have available for export—as foreign currency earners or for purposes of economic warfare—substantial amounts for Europe, the United States, and Japan.

By 1980, for example, the USSR could be producing some 12.6 Tcf/year, importing (for convenience' sake) some 0.5 Tcf per year and exporting perhaps 1.5 Tcf per year to Europe—which by then could represent 10 percent of Europe's consumption. LNG exports—not before the early 1980s—could mean 0.7 Tcf per year from West Siberia to the

U.S. East Coast. From East Siberia to Japan and the U.S. West Coast there could be an unknown, but presumably important, quantity of gas; these are all "potentials" based upon scant information. There is little doubt, however, that the USSR could and, therefore, may be among the leading exporters of natural gas by the end of the next decade should it want to be and should it master the very real problems of priorities, investment, technology, and logistics. Currently, Soviet natural gas exports are to Austria and Germany (of the West), but additional agreements have been or may be concluded with Italy, France, Finland, and Switzerland; Japan remains a possibility.

Summary

The prospect for very large amounts of gas in world trade, for the next several decades at least, depends crucially on whether the reserves of the Middle East and the USSR are available. Even if they are, the security issues raised by natural gas imports will continue to pose—or ought to pose— serious doubts whether energy import–dependent nations should further compound their already complex situations by natural gas imports from these particular sources.

5
Nuclear Energy, 1977-2000

Nuclear fission is the next great development in civil energy. Technology itself—that is, technological reasons associated with present-generation reactors—places no limits on the ushering in of the nuclear fission era, the more extensive use of nuclear energy (in quantities sufficient to displace oil), and the degree and speed with which nuclear energy spreads to all nations. Nuclear fission technology and fabrication capabilities are now widespread in the industrial states and are thus available for sale to the rest of the world. Instead, questions of comparative and absolute costs and problems associated with the fuel cycle—including safety—will determine the rate of growth (and hence the contribution of nuclear energy to total energy supply) and the scope (the numbers of countries participating) of the nuclear era.

The nuclear fission era, with its own set of geopolitical factors, will overlap the declining use of oil as fuel in the closing years of this century and for decades into the next. When technical advances allow the utilization of solar energy for generating larger amounts of electricity at competitive cost, the nuclear era will in turn fade. But for the purposes of this analysis, energy derived from nuclear fission will be of increasing importance—in terms of its contribution to energy supply and its rapid global spread—for the remainder of the century and for some period into the next.

All this is not meant to suggest that there are no unresolved issues affecting the further development of conven-

Table 15. Projected Nuclear Power (in gigawatts electrical)

Country/region	1973 OECD forecast			1980		1985		1990		2000	
	1980	1985	1990	1	2	1	2	1	2	1	2
United States	132	280	580	77	82	185	205	339	385	805	1,000
Canada	7	15	31	7	7	18	18	41	41	115	115
European Community	57	134	283	59	56	146	159	285	292	633	623
Japan	32	60	100	16	17	50	49	90	84	190	157
Others	35	78	146	31	32	64	99	100	202	215	585
Free world	263	567	1,068	190	194	463	530	855	1,004	1,958	2,480

Note: 1. Edison Electric Institute; 2. OECD/IAEA, "Uranium—Resources, Production and Demand."

tional nuclear energy. Clearly, there are such issues, and the initial great hopes for an early and accelerated use of nuclear energy are now being soberly reassessed. The most recent forecasts of nuclear power expansion have been scaled down (see table 15) from past estimates. For these reasons, it is now generally believed that conventional nuclear energy can only begin to make a substantial contribution to energy supply in the late 1980s and early 1990s.

The general economic slowdown in 1975 was undoubtedly responsible for some of the deceleration in planned nuclear energy programs. But this should not obscure the fact that problems relating more directly to nuclear power itself, including public opposition, will prevent conventional nuclear energy from making the fullest, technologically possible contribution to total energy supply. Any shortfall in available nuclear energy will have to be met from increased oil imports.

The greater utilization of conventional nuclear energy seemed to receive a major boost when, in 1973, the vulnerability of the industrialized countries to oil supply disruptions became apparent. Developed countries countered, in part, with threats to accelerate the development of nuclear energy. In this way, it was argued, Western technology, reflected in the development of nuclear power, could displace oil from some of its uses and reduce dependence on imported energy sources. Today, the magnitude and actual materialization of these benefits from nuclear energy, in the intermediate term at least, is in doubt. In the past two years, orders for nuclear reactors have been canceled, and orders for a great many others have been delayed for up to five years. Currently, the United States has 58 nuclear power plants on line. Projected plants for 1985 are now 170, according to one source; in 1974 it was anticipated that there would be 209 in 1985.

Such postponements reinforce a point made repeatedly in this report: progress in energy matters depends upon the timely taking of closely interrelated steps. This, in turn, urgently requires government involvement and coordination in the development of adequate energy supply. In no area is this clearer than in the nuclear field, where costs are high, the industry is not integrated (with strong possibilities for leads and lags in the complex and interrelated aspects of the nuclear fuel cycle), and the security implications are enormous.

If there is no government involvement, a continuation of present trends also suggests a strong possibility that a scarce and precious resource—uranium—may be utilized in a manner considerably less efficient than is prudent. This will be the case if present-generation reactors based on present enrichment technology continue to dominate nuclear energy throughout the world. Any nation whose energy development entails an inefficient use of its uranium resources has no guarantee that adequate imports will be available to it in the future, a matter to be discussed later in this chapter.

Further into the future, the refinement and commercialization of the "breeder" reactor (which, over a twenty-year period, creates more fuel than it uses) and the later development of nuclear fusion (based on almost limitless supplies of deuterium) represent changes still to come in the nuclear era. Moreover, these developments could largely free countries from the constraints of the geopolitics of energy. The use of the breeder reactor would result in a profound reassessment of the requirement for enriched uranium and thus for the ore itself, substantially freeing nations from the constraints of resource scarcity and the tyranny of the location of energy resources beyond their borders. Fusion should free nations from these constraints.

Nuclear Energy as a Substitute for Oil

As presently envisaged, nuclear energy will be used over-whelmingly in the generation of electricity, progressively dis-placing conventional fossil fuel electricity generation. In 1975, 79 percent of the electricity produced both in the United States and Europe of the Nine and 83 percent of Japanese electricity was derived from conventional thermal sources.

The potential magnitude of the nuclear contribution to total energy supply is dependent on the growth in electricity demand. Although it could produce heat for various indus-trial purposes as well, nuclear energy (particularly nuclear energy produced by the current Light Water Reactor [LWR]) is primarily applicable to electrical generation, leaving fossil fuels dominant in other areas, such as transportation and petrochemical products. The quantity of fossil fuels used for current and possible future generation of electrical energy represents the maximum amount that might be displaced by nuclear energy, at least for as long as nuclear power is derived from large units.

Historically, growth in electricity demand has been very rapid—approximately twice the rate of energy consumption as a whole. Despite recent slowdowns, which will not alter historic patterns, it is anticipated that electricity will provide an increasing proportion of Free World total energy supply. Currently, electricity represents 15 percent of U.S. energy consumption and 9 percent of the gross inland energy con-sumption of Europe of the Nine. While nuclear power may contribute no more than 14 percent of Free World primary energy supply in 1985 and some 24 percent in 1990, elec-tricity may represent 30 percent of Free World energy supply in 1985 and a higher proportion thereafter.

The versatility of electrical energy is widely acknowledged: it can be produced from oil, gas, coal, nuclear, water, geo-thermal, and solar energy sources. However, electricity can-not now be "stored" in the general sense, it is expensive to generate and transport, and its production consumes a great deal of energy (but it is more efficient in end-use than conventional fuels). Electricity concentrates production of pollutants in a single, highly visible plant (while electricity itself is a clean energy source). Siting, environmental problems, and the financial difficulties confronting the utility companies will affect the supply of electricity, but the upward trend of electricity in total energy supply is assured.

Nuclear power will be called upon to meet not only electricity demand growth but also to compensate and replace obsolete thermal generating plants. By 1985, 26-30 percent of U.S. electrical generating capacity may be derived from nuclear energy; currently, nuclear energy represents only 5 percent of total electrical generating capacity. In Japan, a similar proportion of electricity will be generated from nuclear power plants by 1985. The EC member countries estimate that 45 percent of their electrical energy will come from nuclear sources by 1985 (compared to 6 percent at present). In Germany, 40 percent of electrical requirements will derive from nuclear generated electricity in 1985; in Italy, this figure will reach 50 percent in 1985 and 80 percent in the 1990s; and the comparable 1985 figure for the United Kingdom is 25 percent. In light of the recent deceleration in nuclear programs, these forecasts are probably optimistic, but they are indicative of the general trend.

In terms of potential savings of conventional energy sources, a conservative forecast of nuclear energy possibilities suggests that nuclear power could displace 1.6 billion barrels of oil equivalent in 1980 and 3.5 billion barrels in 1985. Divided among the industrial nations, in various proportions, these figures suggest that the amounts of oil used for electricity generation, and therefore susceptible to displacement by nuclear energy, are not large. In the United States, only 554 million barrels of oil a year (or 9 percent of U.S. petroleum inputs to all sectors) are used for electricity generation.

Since an increasing electricity demand is likely, however, savings of oil, oil that in the absence of nuclear power would have been required for electricity generation, may be substantial. Individual nations may decide that the development of nuclear energy is worth the very high cost involved. This will be particularly true if domestic uranium resources and enrichment facilities can substitute for imported oil. Thus by concentrating only on the quantity of oil displaced, one misses the point: to the extent that nuclear energy represents an alternative domestic energy source or a geographic diversification of energy sources, its value in terms of national security and freedom of action far exceeds the value implied by the oil displacement numbers alone.

Energy "Independence"

The role that nuclear energy may play in reducing supply uncertainties will be limited in large measure by issues affecting the nuclear fuel cycle. Current nuclear technology is based primarily on the uranium fuel cycle. Uranium ore is mined, milled, and refined to produce uranium concentrates, U_3O_8; this is then converted to uranium hexafluoride (UF_6) to provide feed for uranium fuel, which is then fabricated into nuclear fuel. In fabrication, enriched uranium is pelletized, encapsulated in rods, and assembled into fuel elements. The fuel is then loaded into reactors, and the heat of the fission process is utilized in electricity generation. Spent nuclear fuel may be reprocessed to recover the remaining fissionable uranium and plutonium. Radioactive wastes produced in the process are then permanently stored.

All aspects of this nuclear fuel cycle are interdependent, i.e., developments in any particular aspect of the fuel cycle will have implications for the rest. Because the different steps are interrelated and because a large proportion of the steps are under governmental control while other steps are in the hands of private enterprise, at least in the United States, the potential for leads and lags and the development of bottlenecks is unusually great. Reactor technology is proved and commercially available. But it is not certain that all the necessary supporting functions will be available for the optimum use of uranium or available in a manner that encourages the maximum development of nuclear energy.

Given the very long lead times involved in the development of nuclear energy, it is possible to be somewhat more confident about the nuclear situation in 1985 than for other energy sources:

- From exploration to production of uranium   8-10 years
    - To open mine: 3 years
    - To build mill: 2 years
- To establish a conversion plant   4 years
- To build a fabrication plant   5 years
- To design, construct, license, test, and put a new enrichment plant into production   8 years
- To construct and begin generation from a nuclear reactor   7-10 years
- To construct a reprocessing plant   10 years

Uranium Ore

Uranium reserves are believed to be concentrated in four countries: the United States, Canada, Australia, and South Africa (where uranium is presently a by-product of gold mining). The cyclical nature of uranium demand to date has resulted in sporadic exploration and incomplete delineation of reserves; therefore, the usual uncertainty regarding all raw material reserves plagues uranium estimates as well. However, these four countries will continue to account for a major portion of uranium reserves and production for the next fifteen years at least, and probably even longer.

In the absence of intensive uranium exploration and development, constraints on nuclear energy developments caused by a scarcity of low-cost uranium reserves could emerge in the early 1980s. Moreover, because of the heavy capital investment costs involved, reactors and nuclear power stations for which an adequate and continuous supply of fuel for twenty years of operation is not guaranteed may simply not be built.

Table 16. Free World: Estimated Uranium Resources
as of Jan. 1, 1975 (thousand metric tons)

	Under $15/lb U₃O₈ production ccst				$15 to $30/lb U₃O₈ production cost			
	Reasonably assured	Per-cent	Additional	Per-cent	Reasonably assured	Per-cent	Additional	Per-cent
United States	320	30	500	50	134	18	312	46
Canada	144	13	324	32	22	3	95	14
Australia	243	23	80	8				
South Africa	186	17	6		90	12	68	10
Subtotal	893	83	910	91	246	33	475	70
Other	187	17	90	9	484	66	205	30
Total	1,080		1,000		730		680	

Source: OECD/IAEA, "Uranium—Resources, Production and Demand."

Comparing uranium production to world demand for uranium thus confirms the prospects of shortages as early as 1980-1985. Producing capacity, even if additional reserves are discovered, could fall short of demand sometime after the early 1980s if no additions are initiated immediately. Moreover, 1985 cumulative demand represents 18-20 percent of total uranium resources ("reasonably assured" plus "additional" at under $15/lb and $15-30/lb); by 1990, cumulative demand will have accounted for 37-43 percent of these same reserves.

Table 17. World Uranium Producing Capacities (thousand metric tons)

	1974	Percent	1975	Percent	1980	Percent	1985	Percent
United States	14	56	12	46	25	42	40	46
Canada	5	20	7	27	10	17	12	14
Australia					3	5	5	6
South Africa	3	12	3	12	11	18	14	16
Subtotal	22	88	22	85	49	82	71	82
Others	3	12	4	15	11	18	16	18
Total	25		26		60		87	

Uranium demand, however, is not determined solely by the demand for nuclear energy per se or by the demand for electrical energy, although clearly these are important determinants. Recycling, as indicated in the table, could reduce uranium requirements. (There are no commercial reprocessing facilities in the United States, and President Carter has announced that there will be no federal financial assistance to any such projects, thereby almost certainly guaranteeing that there will not be any commercial reprocessing facilities in the United States.) Reactor type and size also have a bearing on natural uranium requirements. The Light Water Reactor, which is technologically proved and the likely dominant reactor type for the rest of the century, uses more uranium less efficiently than some of the other existing reactor types. The type and amount of enrichment also affect demand for natural uranium, and the LWR requires highly enriched uranium.

Table 18. World Demand for U_3O_8 (thousand metric tons)

	Edison Electric		OECD/IAEA	
	No recycle	Recycle	High*	Low†
1975	23	23	18	18
1980	61	56	53	48
1985	115	99	101	82
1990	191	153	168	130
2000	336	281	313	326

*Assuming no plutonium recycle.
†Assuming some constraint in electricity demand growth and plutonium recycle as from 1981.

Further uncertainty derives from efforts to develop and commercialize the breeder reactor. Whether the "breeder" comes into use may be a highly significant determinant of the adequacy of uranium reserves and of planned enrichment facilities. If, as some assert, European "breeder" technology (especially that of France) is ahead of United States technology and given President Carter's recent decision, the effect of which is to postpone development of the breeder in the United States, some difference in comparative uranium-supply security may emerge, but not to any appreciable extent before the 1990s. Widespread use of a "breeder" could extend the life of uranium resources by sixty years or more.

The possibilities for world trade in uranium ore do not appear to be great. Producers can be expected to satisfy domestic requirements before considering export. In addition, the value of enriched uranium is about three times the value of

Table 19. World Cumulative Demand for U_3O_8
(Uranium concentrate, thousand metric tons) *

	No recycle	Recycle
1975	23	23
1980	232	218
1985	687	619
1990	1, 487	1, 281
2000	4, 226	3, 532

*Tails assay 0.3 pct.; 72 pct. equilibrium capacity factor.
Source: Edison Electric as quoted in the Atlantic Council.

natural uranium. Producers, therefore, are likely to delay exports until a national enrichment capability is achieved. (In the United States, imports of uranium were banned until 1977, when an incremental lifting of the import prohibition took place.) The implications of an emphasis on exporting not the ore, but the enriched fuel itself, raise the same security issues that would face countries dependent on refineries located in oil-exporting nations rather than on the latters' crude alone.

The fact that no OPEC or OAPEC state is now a large provider does not automatically assure a supplemental supply. Although this is of interest, obviously, it is possible that one or two or conceivably even three (South Africa, Australia, and Canada) could wish to employ their resource position to attain some economic or political objective. While it seems nearly inconceivable to Americans that such a combination could be raised against them, the possibility exists. There is also no necessary identity between the interest of producers and consumers, and it would not take a political objective for producers to deny access to consumers except on the former's terms. In the uranium producers group, there may be the beginning of concerted producing-country policies and activities, not unlike OPEC's efforts.

We do not yet know enough of the location of substantial uranium reserves elsewhere, but aside from North America and the others previously mentioned, Gabon, Niger, Algeria, Pakistan, Brazil, and others indicate that some, possibly large, reserves will be located in the LDCs.

Table 20. Free World: Estimated Nuclear Capacity
by Reactor Type (GWe)

	1980	Percent	1985	Percent	1990	Percent	2000	Percent
Light water	170	88.0	475	90	860	86	1,842	75
Steam-generating heavy water			5	1	21	2	77	3
Advanced gas-cooled	6	3	6	1	6		5	
Candu—heavy water	10	5	27	5	68	7	183	
High temperature	1	.5	8	1	30	7	136	7
Gas-cooled, graphite-moderated	6	3	4	1	3			5
Fast breeder	1	.5	5	1	16	2	237	10
Total	194		530		1,004		2,480	

Source: OECD/IAEA, "Uranium—Resource, Production, and Demand."

Enrichment

Enrichment capacity is now a potential bottleneck in the continued development of nuclear energy. In this case, and for a time, the U.S. government provides about 95 percent of Free World enrichment capacity at three plants located in the continental United States. It is anticipated that even with the expansion of these facilities enrichment capacity could be fully saturated in the next ten years, suggesting that for another decade it may be the United States, through its enrichment capacity, that should continue to be greatly influential in nuclear development. However, it is also a "wasting" asset, since the Europeans and others (i.e., South Africa) are working in this area of fuel supply.

Required enrichment capacity is also determined by factors beyond the demand for nuclear energy. Reactor type, the availability of recycling facilities, and the tails assay (the amount of U^{235} left in the depleted portion of uranium feed) of the enrichment process itself will affect the level of need for enrichment capacity.

Most recently, the USSR has begun supplying natural and enriched uranium to Western Europe, a development that should redouble concern over the extent to which the USSR

Table 21. Present and Projected Enrichment Capacity
(thousand tonnes SW per year)[1]

	1980	1985	1988
United Kingdom	0.4	0.4	(0.4)
United States [2]	27.7	44.9	50.8
URENCO [3]	1.0	10.0	10.0
EURODIF	6.5	10.8	10.8
EURODIF II [4]		6.0	9.0
UCOR (South Africa)			5.0
Total capacity	35.6	72.1	86.0

[1] SW (separative work)—effort involved in various enrichment techniques is expressed in terms of SW units.
[2] Includes new plant of 17,200,000, tons SW per year for 1985 and 23,000,000 tons SW per year in 1988.
[3] Capacity will be increased according to requirements.
[4] Under consideration.

Source: OECD—Note that adequate enrichment capacity for 1985 and beyond is dependent on projects currently only planned.

may, in this field, be acquiring another point of energy leverage against NATO allies. The Europeans also have plans for building an independent enrichment capability; the Eurodif project (France, Italy, Belgium, Spain, and Iran) envisions the construction of an enrichment facility in France to be operating at full capacity (10.8 million separative work units [SWU]) in 1981. The URENCO project (the Netherlands, Germany, and the United Kingdom) is expected to attain a capacity of 10 million SWU/year by 1985. In 1985, however, the United States may still retain over 60 percent of total world enrichment capacity.

Safeguards

Adequate safeguards against possible accidents in the manufacture and use of nuclear energy are of extreme importance both because of the damage and contamination resulting from a nuclear accident and because of the implications of an accident for public attitudes. In the event of a major accident, it is likely that in many countries, public reaction would delay still further the fullest potential value to be derived from the application of nuclear energy.

Table 22. Separative Work Units Required
(thousand tonnes SW, 0.25 pct tails assay)

	Annual demand			
	High estimate		Low estimate	
	Pu recycle	No pu recycle	Pu recycle	No Pu recycle
1980	31	31	28	28
1985	58	65	51	57
1990	98	112	84	95

Source: OECD.

In addition, the danger of even more rapid nuclear weapons proliferation, as a by-product of the use of nuclear energy, requires that adequate safeguards against such an eventuality be developed. Weapons-grade fuel can be produced in
the recycling process. This largely explains U.S. resistance to
reprocessing; on the other hand, Europeans and LDCs, in a
much less favorable uranium resource position, take for
granted the necessity of reprocessing.

The enrichment process itself can produce weapons-grade
uranium. The regional enrichment-processing centers proposed by the United States can be seen as serving two ends:
first, it is an attempt to prevent fuel services from becoming
manipulable and tools for the pursuit of political objectives
by the country selling such services; and, second, it is an attempt to introduce a safeguard against the diversion of enriched uranium from peaceful uses to nuclear weapons.

It is difficult to escape the conclusion that the further development of nuclear power, particularly (but certainly not
exclusively) in LDCs, not many of which will be able and
willing to bear the cost of nuclear development, and the export of nuclear reactors and fuel cycle technology (which the
United States will remain powerless to stop) involve a significant risk of accelerating the proliferation of nuclear weapons.
It is highly doubtful that the Nonproliferation Treaty will
stop any nation from pursuing its own conception of its national interest in this extremely sensitive area. More recent
developments in the nuclear Suppliers' Club, which consists

of all the major industrialized exporters of nuclear facilities and technology (including the Soviet Union), and the obvious importance President Carter attaches to nonproliferation of nuclear weapons suggest a greater possibility that additional international safeguards will be agreed upon. But time is short, and the approach—cooperation among suppliers alone —may be inadequate. Nuclear technology is increasingly widespread, and some of the non-nuclear-weapons countries are cooperating among themselves (e.g., Iran–South Africa, Israel–South Africa).

The recent deterioration in U.S.-Brazilian relations— caused by unsuccessful U.S. pressure on West Germany to cancel the sale of reprocessing and enrichment facilities to Brazil—is only one indication of the difficulties involved in international nuclear relations. It will be difficult to find common ground among: (1) those countries seeking nuclear facilities, including domestic reprocessing and enrichment capacity, in order to achieve some greater degree of energy independence; (2) those countries in which the viability of the domestic nuclear industry depends on the ability to export nuclear technology; and (3) those countries that see in the spread of reprocessing and enrichment facilities an unacceptable risk of nuclear weapons proliferation.

Conclusion

There is a strong possibility that: (1) uranium resources will not be utilized in the most efficient manner, and (2) the future development of nuclear energy will be impeded by bottlenecks in several phases of the nuclear fuel process. There is no inherent or technological reason why these two events must occur, but in the absence of government involvement and coordination there is little likelihood that they will be avoided.

Competition for natural and enriched uranium may parody developments in oil. Europe and Japan do not escape energy import dependence via the conventional nuclear energy route. However, if enrichment capacities permit, a prudent measure for the United States and certainly for the Europeans and Japanese involves the stockpiling of enriched fuel in a fabricated state to protect against a power breakdown if supply were severed and a country were deprived for a long period of time. The increasing supply of enriched uranium from the Soviet Union to Western Europe deserves watching.

Moreover, in spite of its associated problems of wastes and plutonium, the emplacement of breeder reactors would be another long-range necessity to obviate reliance upon foreign ore and enrichment capacity until the coming of nuclear fusion and, even further in the future, solar power. While the United States has apparently opted against the breeder, it may not be realistic to expect countries with fewer nuclear resources to do likewise without any quid pro quo, e.g., enhanced cooperation in the nuclear fusion area.

Thus, development of nuclear energy as currently envisaged may not result in energy independence. However, with recycling, the breeder, and additional technological processes that extend the life of uranium reserves, some greater degree of independence from energy imports will develop, and this is what the Europeans and Japanese are counting on. Eventually, fusion may significantly reduce the energy import dependence of the industrialized states but not before the first or second decades of the next century.

6

Import Dependence of Industrial States

The continuing, general dependence upon the Middle East is evident. Earlier discussion has stressed the varying degrees of dependence upon oil and upon the Middle East among Europeans, Japanese, and Americans; more specifically, there are significant differences in the degrees of dependence upon particular Middle East sources. Current import figures make this clear.

Before presenting these data, a warning is necessary. Degrees of dependence upon oil are all "one-directional," that is, oil continues to be an ever more important energy source for most nations. Similarly, for the next fifteen to twenty-five years, it is not likely that the percentage of imports in national energy budgets will be greatly diminished. Even if the percentage represented by oil is smaller, volumetric requirements will increase. Some 50 million barrels a day were consumed in 1975; by 1990, even with efforts by governments to reduce the percentage represented by oil in their energy balance (some of which will be successful), we anticipate that close to 80 million barrels of oil will be consumed.

In addition, there will be the usual shifts in reliance upon particular sources for the ordinary commercial, seasonal, and marketing reasons, for competitive purposes, or because one source or another may seem more or less reliable. Therefore, one can only observe *annual* trends, not conclude too much from *quarterly* import figures, and never conclude from monthly figures. Finally, shifts in sources cannot be made

Table 23. Estimated Percent of Imports (Crude and Product)
Traced to Original Source, 1976

A. Arab

	Saudi Arabia	Kuwait	Libya	Iraq	UAE	Algeria	Total
U.S.	18		7		4	6	36
Canada	15		3	4	2	2	26
Japan	33	9	1	2	10		55
Western Europe	26	5	7	10	5	4	57
U.K.	18	11	3	5	4		42
Germany	13	1	15	1	5	7	43
Italy	23		11	14		3	50
France	34	3	2	12	9	4	65
Spain	38	6	9	9		3	65
Netherlands	25	8		2	8		44

B. Non-Arab

	Iran	Venezuela	Indonesia	Nigeria	Canada
U.S.	7	13	8	15	8
Canada	19	36		4	
Japan	18		12		
Western Europe	16	2		6	
U.K.	19	2		4	
Germany	14	1		6	
Italy	13	1			
France	11	1		6	
Spain	18	2			
Netherlands	23	1		14	

Source: Derived from Central Intelligence Agency, *International Oil Developments.* Washington, D.C., May 1977, pp. 5-9.

peremptorily, as it were; the management of worldwide crude slates and exchanges, the logistics, and refinery functions must all be anticipated and interwoven.

The following observations are important: (1) the United States received a lower (but rapidly increasing) percentage of its oil imports, by a substantial degree, from *Saudi Arabia* than either Western Europe (as a whole) or Japan did; (2) the United States also received a lower percentage of its oil imports from *Iran* than any other major importing, industrial state did; (3) the United States received a higher percentage of its oil imports from *Nigeria* than any other major importing industrial state did; with the single exception of Canada, the United States received a far higher percentage of its oil imports from *Venezuela* than any other of the listed states did; and (4) the United States received, in 1976, some 36 percent of its oil imports from Arab countries (up from 28 percent in 1975), while Western Europe got nearly 60 pecent and Japan 55 percent of their oil imports from Arab coun- tries (if Arab imports to the Caribbean refineries for final destination in the United States are included, U.S. depen- dence on Arab sources increases to 38 percent, but remains substantially lower than the figures for Europe or Japan).

If we look at exports from the *perspective of the pro- ducing countries,* the exceptional importance of the *Euro- pean* or *Japanese* markets to the oil-exporting countries is evident. Although 1976 figures would show the United States taking a higher percentage of Saudi oil than the figures shown here, the general point of the greater importance of the Western European and Japanese oil market remains valid.

Only in the case of Nigeria and Canada does the United States dominate a producing nations' export market. (Vene- zuelan exports to the United States have occupied a more im- portant place in its trade; the lower percent largely reflects the recession's drop in fuel oil demand.) For OAPEC as a whole, the United States receives 10 percent of members' exports, Western Europe gets 42 percent, and Japan 15 per- cent. For OPEC as a whole, the United States receives 18 percent of their exports, Western Europe receives 40 percent, and Japan 16 percent. Put in mmb/d, OAPEC exports some 1.8 to the United States, 7.5 to Western Europe, and 2.7 to Japan; OPEC exports some 5.4 to the United States, 12 to Western Europe, and 5 to Japan.

These data emphasize the underlying oil market factors that help explain the caution with which Western Europe and Japan must consider U.S. "initiatives" vis-à-vis OPEC. There can scarcely be any meaningful comparison between Europe and Japan, and the United States, in this respect, that is, their comparative importance to oil in world trade—OPEC oil or, even more important, oil from the region of the Gulf.

The nuclear section of this report suggested how likely it is that the industrial nations will depend not only on oil im- ports but on uranium ore imports as well. Thus, the energy concerns of these states will be also for adequate and contin-

Background to Policy

Table 24. Estimated Percent of Oil Exports
to Listed Importing Nations, 1975

	MB/D total	United States	Canada	Japan	Western Europe
From Arab countries:					
Saudi Arabia	7,080	10	3	19	47
Kuwait	2,100		2	20	37
Libya	1,520	14		4	48
Iraq	2,250		1	4	41
UAE	1,700	7	3	24	45
Algeria	930	28			54
From non-Arab countries:					
Iran	5,350	5	4	21	36
Venezuela	2,350	16	11		11
Indonesia	1,310	28		40	
Nigeria	1,790	41		4	41
Canada	1,460	100			

uous supply of this other and newer energy source. The only significant difference to be noted at this time is that the present OPEC members do not possess substantial uranium reserves (with the possible exception of Nigeria). It is equally important to note, however, that with the exception of the known large possessors of uranium reserves—South Africa, Canada, the United States, Australia—the sources of additional, and probably needed uranium, lie within the LDCs.

Moreover, imported natural gas is likely to be an important security factor to energy-deficient states, especially in Europe and Japan, as Soviet and Middle East supplies become more available. The major industrial countries can expect their import dependence, including foreign sources of oil, gas, and uranium and enriched uranium, to persist and increase. For very few countries, perhaps only a handful, but including the Soviet Union, is energy autarky feasible in the decades ahead.

7

Governments and Enterprises in International Energy

Role of Governments in International Oil Supply

A major actor in the geopolitics of energy is the giant international oil company, whose supply system and management daily moves enormous volumes of oil around the world —perhaps as much as two and a quarter billion barrels of crude and product. Their ability to deploy oil on such a scale makes them vital to producers and to consumers. Yet while this function remains, the international oil company has been profoundly changed in recent years, and the nature and consequence of the change is an important ingredient in energy policy and programs.

The success of the private international oil industry in making oil the primary source of fuel for the industrial world was the very factor that brought governments of all stripes and sizes into their act. Once oil had become a commodity vital to their economies, decisions affecting national energy interest would not be left only to the commercial instincts of the private commercial or trading sector. Similarly, when oil became crucial to the revenues of producing states, decisions as to volumes and prices could no longer be left to the judgment of oil companies alone.

However, even with that diminished role—particularly in determining the volumes and prices for crude when there is no alternative—it was necessary for consumer governments to undertake larger roles in oil partly to assure continuity of supply. As governments have done so, and the process is still very much under way, a very broad range of government responsibilities became involved. Commercial considerations, which in the past determined the commitments of the oil industry, are no longer central. "Oil" has become enmeshed in a number of other national interests and objectives that complicate "access." Of course, this is true for energy-deficient or importing states as well as for the oil-exporting countries.

Governments' role in oil accelerated with the collapse of the Western empire, chiefly because the international oil majors were linked with the colonial-capitalist system. These companies lacked the foresight, imagination, and initiative to forge timely new relationships with the recently independent and highly nationalistic governments with which they had to deal. Yet the point must also be made that for over a quarter-century, in a number of cases, they had maintained a concession system as the basis for an oil exploitation that was enormously beneficial to them and also to consumers. When the domestic and international ramifications came to the fore, the "politics of oil" took over, and the concession system was the first casualty.

It was really not a case of great changes coming at great speed, for there had been many warnings in the quarter-century after World War II: e.g., Iran, Libya, Iraq, Venezuela, Indonesia, Algeria, and Saudi Arabia. Adaptation might have been impossible for most of the companies in any case. The "majors" saw themselves beset by challenges and dangers but hoped that with time and fortitude all would be right again, or very nearly, or enough so. The attempts that were made to adjust came very largely from the "nonmajors," or so-called independents, who sought an advantage for themselves in accommodating to change by altering traditional concession arrangements and by dividing the profits from oil on terms increasingly more favorable to the producing governments.

The elevation of oil to its current level of governmental concern may make for less difficulties in most industrial nations than it does in the United States. It has long been the case, for much of Europe and for Japan, that major commercial enterprises function within a system in which government concerns and corporate undertakings are related. In fact, for much of Western history great commercial enterprises have often been conceived and sponsored by government. It is only in relatively recent times, and primarily in the United States, that "government" and "commerce" have been seen as separate and even adversary forces. With the politicization of oil an accomplished fact, European and Japanese societies may have an advantage over the United States, which will be wrestling with the question of government-and-industry or government-versus-industry for many years to come.

Basically, however, the justification for some forms of government involvement is that international energy supply is now very largely under the control of states, not commercial enterprises, and that the supply available may therefore be distributed with political purposes in mind. Moreover, with the vulnerability of energy-deficient states to shortages, contrived or otherwise, only government can insure that all possible measures are taken to limit the damage that producing governments can cause. Earlier, when the principal purveyors of oil in domestic and world trade were usually the affiliates of giant international companies with headquarters in London and New York, other governments began to involve themselves to put the affiliates under public supervision. Since there was no adequate private industry to move into oil and compete at home with the large foreign oil companies, there was no alternative to government supervision or participation.

In addition to government attention to security of supply, the enormous expense of developing alternative energy sources and the cost of some energy research now seem to require government direction and funding, directly or through subsidization. There are differences of opinion over the extent to which government involvement is needed, but the general trend toward a larger role is unarguable—as is the need for government attention to legislative or bureaucratic impediments to energy development.

Finally, energy objectives call for development and infrastructure on a larger scale, and time is pressing. Commercial enterprises alone cannot be expected to deal with very large-scale national energy requirements, and they have themselves made this point.

For all these reasons, some government presence is necessary and inevitable. Whether its involvement will improve the provision of energy is not so certain.

The International Oil Industry

While it is inevitable that governments will increase their involvement in energy supply, the international majors—for as long as consumer and producer governments pursue policies that permit them to be important in the search for and supply of oil—can expect to remain essential. On the other hand, if governments give precedence to nonmajors, or if they discriminate in favor of government oil entities or otherwise give preference to national oil companies, then the diminished role for the international majors will have its inevitable effects upon the companies' own interest in keeping in the game.

In judging whether preference should be given to companies other than the international majors, governments have paid insufficient attention to efficiency of supply, which these large companies have come to represent and which is an essential ingredient in the provision of energy. Insufficient attention has also been given to the point that if governments can limit their attention to setting the appropriate framework for corporate risk taking and investment to insure that the necessary size and diversity of energy efforts are, in fact, undertaken, then the energy costs are not a direct charge upon the government's national budget—a not inconsiderable point.

Will there be sufficient incentive for the majors to be interested in remaining in energy? If there is, there can be little doubt that the assets that they possess and that are not readily duplicable—managerial and otherwise—can be employed in their private and in the general interest. In the past, these companies were not "buffers" between producers and consumers, as they often like to say, but the critical link between the two, agents with great interests in both production and consumption, able to balance supply and demand with exceptional skill and efficiency. Under appropriate government-established conditions there can be an identity of interest between the private and public interest.

The international oil companies have seen their role diminished essentially by two forces: (1) the actions of producer governments, which have very largely but not completely removed the companies from decisions affecting crude volumes and prices, and (2) the rise in the numbers and importance of private and governmental oil companies, which are usually nonintegrated, but which have progressively obtained an increasing share of the market in world oil trade.

Before the producers quickly took control of the terms on which they make their oil available, there had been a steady increase in the share of functions performed by nonmajors. There have been many signals that the process will probably accelerate. Whether this trend is beneficial is relevant to the objective of supply security.

In 1961, governments controlled about 8 percent of crude production. The nonmajors likewise controlled about 8 percent. The majors handled nearly 83 percent of the world's oil. By 1972 (the year before producer governments took over), the share of governments had risen to some 10 percent, the role of nonmajors to about 18 percent, and the share of the majors had declined to about 72 percent. The role of government is today at least 70 percent with the balance a nearly indefinable mix of nonmajors and majors. In refining, the role of majors slipped from nearly 70 percent in 1961 to about 54 percent in 1972 (with nonmajors acquiring some 25 percent in 1972). In marketing, the majors went from about 64 percent in 1961 to nearly 52 percent in 1972, while nonmajors captured some 27 percent by 1972 and governments had acquired about 21 percent. These trends have continued.

The most the majors can hope for is to hold their present position; out of it, in time, might evolve opportunities for greater investments. Other opportunities may come from countries determined to explore for indigenous oil. There is also the possibility of roles for these companies, akin to ARAMCO's expanding activities, in producer states. Anything much less than that would surely find an increasing number diversifying their talents and assets into other endeavors.

From the viewpoint of producer governments, it is the ability of majors to move the vast bulk of oil in world trade to the consumers' markets that is now so important. Moreover, in a priod of general oversupply, the majors can make decisions as to supply from which sources, which relieves the governments from having to do so in an OPEC forum. With time, with the universal availability of communications and data processing, consumer or producer government oil companies could move these volumes; but very few, if any, governments today possess the managerial resources needed to cope with the complex supply arrangements inherent in oil moving in world trade. As they acquire the necessary talents, a diminishing role for the international majors would seem to be a natural consequence. But in another function the majors will continue to be better able to mount the kind of immense undertaking represented in the North Sea, on the North Slope of Alaska, and generally offshore. They have no monopoly of such capabilities, but it is still impressive enough to warrant consideration when their future role is discussed.

In these respects—logistics, exploration, and development— the international majors, buffeted as they have been, are still essential and will be essential for many, many years.

Role of Governments and Enterprises in Nuclear Energy

In the nuclear case, the role of government has been preeminent from the outset; it has been the possessor of nuclear technology, the major sponsor of advanced research, and the owner or licensor of the bulk of present enrichment processes. While equipment has been fabricated by both government and private enterprise, there seems little reason to doubt the continued predominance of government everywhere in all parts of the fuel cycle.

The international oil majors—especially Shell, Gulf, and Exxon—have invested with varying success in parts of the fuel cycle, and each has undertaken substantial research programs independently and in concert with government. In no case has any U.S. private enterprise yet obtained a position in all of the fuel cycle. In other countries, such as Germany and France, there appears to be a far closer coordination of effort with private companies. This is in large measure because their governments have adopted clearer guidelines for the support of companies in overseas nuclear contracts.

The United States has been inconsistent and uncertain in its own approach, thus adding to the hesitation of U.S. nuclear enterprises to commit requisite talents and sums to overseas opportunities.

The astronomical rise in research and capital costs; public concern over safety and environmental aspects, including weapons' proliferation; fluctuations in market forecasts for electric power generation; excessive lead times and the unavoidable dependence upon government policies, which are themselves evolving—all combine to limit the interest of private enterprises and to leave to government nearly everywhere the key roles in nuclear energy development. Initiatives for changing these roles have generally faltered, as in the U.S. government's Nuclear Assurance Fuel Act, which was the single most important move to enlist private enterprise in the key enrichment processing function. Only in the design and fabrication of capital equipment, including reactors, will U.S. private enterprise be likely to have an important role.

Because substantial uranium ore imports will certainly be necessary for the industrial nations, the government role in securing access to uranium ore and in the acquisition of enrichment and reprocessing facilities insures the politicization of

nuclear fuel supply in all respects. This is abundantly clear
from President Carter's persistent effort to limit the spread of
nuclear fuel reprocessing so as to lessen the chances of a plu-
tonium-weapons proliferation and disaster. The proposed
slower pace of U.S. attention to the "breeder" is a result.

But for many countries, nuclear energy seems to be their
only escape from oil; for them, lacking the energy resources
of the United States, there is no alternative. From their perspective, there may well be a race for uranium not unlike the
race for oil, so there is little prospect that they will discard,
at the United States' behest, their efforts at reprocessing
technology and the "breeder" reactor, efforts undertaken to
obtain more energy self-sufficiency than would otherwise be
the case. The United States' near-monopoly of the nuclear
fuel enrichment process is fading; scientific advances in fission and fusion warn of a spreading capability that is likely
to spawn additional problems for the nuclear and uranium
suppliers, for which a kind of "nuclear OPEC" is altogether
possible, if with quite different actors.

8
Summary and Overview

Most forecasts of energy supply and demand are based on certain assumptions: (1) a decline in oil demand resulting from higher oil prices; (2) a decline in energy demand arising from deliberate conservation schemes; (3) an expanded indigenous (non-OPEC) production stimulated by higher energy prices; and (4) the development of alternative energy sources, also encouraged by higher oil prices. It is further assumed that future GNP growth rates will fall below previous rates (in part, because of higher energy prices), suggesting some moderation in the growth of energy demand. In addition, the forecasts generally assume the requisite industry investments and positive expressions of government support for conservation and the development of energy resources, through a variety of interrelated policies implemented in timely fashion.

The sensitivity of the forecasts to any change in their basic assumptions is illustrated dramatically in the 1973 OECD study, "Energy Prospects to 1985." In the OECD example, the projection, based on a current dollar price of $9 per barrel of oil, includes very optimistic assumptions about the ability to expand OECD indigenous oil production, the ability to develop alternative energy resources, the oil savings to be derived from conservation, and the decline in oil demand resulting from higher oil prices. Oil imports provide the balancing mechanism between OECD oil supply and OECD

Figure 3. OECD Oil Imports—1985

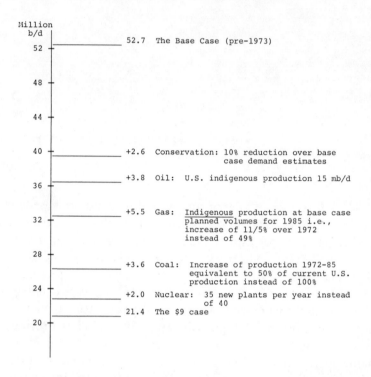

oil demand. If any of the assumptions about alternatives, indigenous production, and conservation prove wrong, they can throw the forecasts off greatly. Figure 3 shows the sensitivity of the forecasts to such errors.

A failure to develop alternative energy sources (coal, gas, nuclear) in the assumed quantities added 11 mmb/d to OECD oil imports in 1985. In all, failure to develop alternatives, failure to find and develop adequate quantities of indigenous energy resources, and failure of conservation efforts added 18 mmb/d to the OECD oil import bill for 1985. While it is unlikely that there will be a total failure in all these directions, a

50 percent shortfall would still be equal to today's total production of Saudi Arabia.

In addition to possible failures in achieving the quantitative requirements, all developments must occur in a timely fashion. Energy supply and demand are determined by a complex and interdependent set of economic and political factors. Failure to complete any particular step in time, or partial but insufficient success in another phase, can trigger a set of consequences that could throw all forecasts off.

In this total energy effort, the critical importance of some form of government coordination at least, and probably involvement, is obvious. The role of industry is also critical. In the absence of an effective national energy policy and its implementation, there is no inherent reason why market forces alone will call forth an industry response in either the necessary direction or with adequate scope and speed. The interrelatedness of all aspects of the energy equation also suggests that ad hoc and isolated government initiatives will not be sufficient to meet national energy policy. The provision of adequate and continuous energy supplies necessary to economic well-being and military security now involves government and industry; anything less is an abnegation of government responsibility. The form of government participation may be a subject for debate; government involvement per se is not.

Given the current uncertainty in government policy and the long lead times required for the full development of any particular energy resource, the energy balance in the industrialized states through 1985 is not expected to differ radically from the current energy supply situation. Oil will retain its dominant place in the world energy balance. Neither gas, nor coal, nor nuclear energy will significantly diminish the import dependence of the free, industrial world. Perhaps the single most important change will begin with nuclear energy, which will account for increasing quantities of electric power generation.

Oil will retain its central place in energy supply, and oil imports will provide the major portion of the oil supply of the

industrialized nations. With scant possibility that any major
oil finds outside the Middle East and the Soviet Union can be
producing at high volumes by 1985, oil imports will continue
to come increasingly from OPEC sources, particularly the oil-
producing countries of the Gulf. Moreover, given the quantity
of oil demanded, "major" finds would have to be huge even
to begin to challenge the dominant position of the Gulf. The
prospect is thus for increasing competition for Middle East
oil; U.S. competitors will include not only NATO allies and
Japan but perhaps the USSR as well.

This situation may well hold into the 1990s, when it is expected that nuclear energy, oil from tar sands, oil from shale,
and coal gasification and liquefaction may be making larger
contributions to energy supply. But until then none of these
developments will eliminate oil's dominant role in total
energy supply.

Figure 4 shows possible changes in the pattern of energy
sources from now until 1990. As in the OECD study, figure 4
assumes reasonably intelligent energy policies and adequate
incentives to industry to invest the requisite financial, technical, and managerial resources either to reduce the role of
oil or intensify the search for indigenous oil resources.

It further assumes substantial success in the discovery of
significant amounts of crude in the major industrial nations
and areas—a prospect that may be too optimistic. The
"savings" referred to come from improved techniques in
energy consumption, better design of equipment, buildings,
transport, new plants replacing old ones, and the like; such
savings are considered to be practically attainable. Moreover,
figure 4 depicts a situation that is general for areas outside
the communist sphere; if realized, it would still imply different
degrees of energy dependence for individual countries.

Three points deserve emphasis. First, even with anticipated
savings and development of alternatives, oil will still provide
approximately 40 percent of total Free World primary energy
supply in 1980. A more recent forecast from the same source
agrees with the Exxon prediction that in 1990 oil may still
account for 50 percent of total Free World energy supply.

Figure 4. Primary Energy Sources of the Free World.

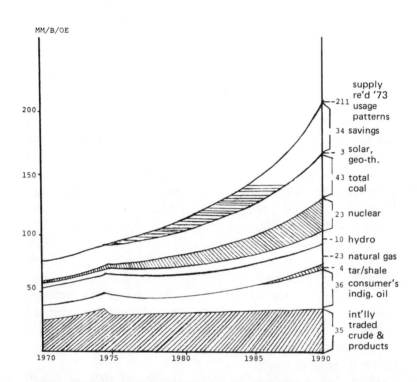

Moreover, even if oil's share in total energy supply should decline, the absolute quantity of oil demanded will increase through 1990.

Second, it must be remembered that only increased oil imports can make up for shortfalls in the development of alternative energy sources and failure to expand indigenous oil production.

Third, in spite of the expansion of nuclear energy, almost 75 percent of the 1990 Free World primary energy supply still comes from conventional energy sources—coal, water, natural gas, and oil. Of these, oil will be by far the most significant energy resource in international trade. Further use of conventional fuels also may not free Europe and Japan from energy import dependence. Additional reliance on gas imports would present the unattractive alternatives of becoming dependent on Soviet gas exports or of multiplying dependence on the Middle East (gas as well as oil).

In addition, developments in nuclear energy, tar sands, and oil shale do not necessarily reduce the energy import dependence of all the industrialized countries equally or at all. Tar sand and oil shale deposits appear to be concentrated in the United States, Canada, and Venezuela. Outside the United States, uranium deposits appear to be concentrated most prominently in Canada, Australia, and South Africa, while enrichment capacity may still be dominated by the United States in the early 1990s. As for the ore itself, while these states are counted among the industrial and advanced developing countries rather than among the OPEC countries, it would be prudent to anticipate that the interests of these producers will not automatically coincide with those of consumers.

Generally, the European and Japanese resource position in energy is clearly less favorable than the U.S. position, and the USSR may be in the most favorable energy resource position of all (over the long term and considering only Soviet domestic requirements).

If little else is done, a continuation of present trends results in the following situation through the 1990s: continued dominance of oil in total energy supply, demand for increasing absolute volumes of oil, the West's and Japan's increasing dependence on oil imports, the increasing importance of Gulf oil, and intensified competition for this oil among the United States, its NATO allies, Japan, and possibly the Soviet Union and People's Republic of China. The varying degrees of our respective dependence upon Gulf exports, with Europe and Japan far more dependent than the United States, will serve to constrain the latter. For the Gulf producers, Europe and Japan's greater importance to them has equal significance.

Throughout this century, and still farther into the future, it is inconceivable that the great industrial areas of today will not still be the main energy consumers and energy importers of tomorrow. Both in the case of oil (for this century at least) and for uranium ore (for as long as present-generation reactors are the principal source of nuclear power), the consumers will be in the "North"; the producers will be in the "South."

A significant change will surely come in the geopolitics of energy as the producers begin ever more to process their raw materials and, probably, to increase their involvement in the marine logistics of supply. But the "bottom line"—the ultimate markets—will still be in the "North," and the trend will generally be to correlate the interests of consumers and producers. The potential exceptions—those cases in which great influence over supply is matched by no pressing need to meet demand levels in international trade—will be very few and even, perhaps, be limited to Saudi Arabia. Nevertheless, there will remain exceedingly important considerations affecting energy supply, considerations that warrant further and more specific mention. These are "location" and "control."

Location and Control as Geopolitical Factors

Oil is indeed where one finds it, but there are additional considerations that make location crucial. The new element in the international oil situation is the combination of location and control in one and the same group of underdeveloped countries. The concentration of oil reserves in a small group of less-developed countries, increasingly more assertive in their international relations, combined with the real need of the industrialized nations for oil and the lack of immediate substitutes, gives the coincidence of location and control a compelling importance.

The oil producers are less-developed countries, and to some extent they share the world view common to most LDCs. Location and control become elements of prestige, instruments of influence and power-bargaining levers to be used to reform or replace the prevailing international eco-

nomic and political system, now dominated principally by
the United States and its allies. This system seems to be ex-
ploitative and designed, perhaps consciously, perhaps inad-
vertently, to secure the interests of the industrialized countries at the expense of the LDCs. From the LDCs' perspective, control over their natural resources that are vital to the
industrial nations holds out the possibility that economic
independence, growth, and development are now attainable.

While the LDC solidarity resulting from a common colonial or neocolonial experience and a common sense of
grievance is real, it must not be overstated. It is opportune
for the oil producers to champion LDC causes in the various
conferences and international organizations involved in the
North-South debate. OPEC can uphold the LDC cause at
little cost to its members by linking the question of access to
adequate and continuous oil supplies at "reasonable" prices
to areas of interest to other LDCs. Moreover, by increasing
their links to other LDCs, the cost of any precipitous action
possibly being considered by external powers is increased.

Non-oil-producing LDCs, suffering enormously under the
burden of higher oil prices, still find the OPEC-OAPEC relationship vital. Disunity would not get them less expensive oil,
and, separated from the oil link, the industrialized countries
would be even more reluctant than they are now to make
concessions to LDC demands for a new economic order.

The skewed location of oil reserves, the success of the oil
producers in securing to themselves the largest share of the
benefits from their natural resources, and the model that this
suggests to other raw material producers raise important
questions of access to raw materials, the terms under which
access is secured, and issues of North-South relations in
general.

From the perspective of the industrialized countries, the
location of oil reserves and the loss of control over them
have compelled a recognition of an uncomfortably more
symmetrical interdependence than was thought to exist.
The acknowledgment of interdependence (indeed, dependence), the need for bargaining, and the uncertainty associ-

ated with dependence and bargaining are unsettling to countries accustomed to assuming that power was their exclusive preserve, that the status quo was the right and natural order of things, and that they had a monopoly on wisdom (and power), which secured the peace.

If the initial U.S. response to the OPEC challenge was a call for solidarity among the industrialized states—a show of force of sorts when the use of force itself has been, perhaps temporarily, rejected—it is now clear that there are differences among industrialized states themselves. Their resource dependence differs; Europe and Japan are in far less favorable resource endowment positions than is the United States. Essentially this means that the United States, with less to risk, has more freedom of action. The differences in resource endowment mean that the European and Japanese perceptions of an emerging world order may be significantly different from that of the United States.

Nations accustomed to declining power and cognizant of their continuing and inescapable dependence may be more willing to deal creatively with interdependence than a nation accustomed to greater independence of action. If the United States can, through its enormous economic and market power, prevent a deterioration in the terms and conditions of access to raw materials, so much the better. But this does not preclude a European-LDC arrangement or new Japanese-LDC relations affecting raw materials—including energy—from which the United States may be effectively excluded, either as a result of its own attitude or even intentionally by other industrial and developing states. Should this occur, the divisive effects upon NATO would be very considerable. Nor are we in a position even to guess intelligently about the capabilities and intentions of the USSR in this changing array of interests.

Location and control of oil reserves has seemingly drawn a line separating an emergent LDC bloc from the industrialized countries. Yet the reality of international politics is far more complex than the superficial division of the world along North-South lines. Developments in North-South relations will certainly have an important bearing on the question of access to raw materials. But the situation is malleable, and the shape of new international relationships is still evolving.

We are in a curious position. The parameters of our energy position are clear, and they are unlikely to change unless national policies of a comprehensive, demanding character are put into force. Yet in no case has a commitment commensurate to the challenge been made—either in the United States, Europe, or Japan. If our energy situation is left to drift, then our vulnerabilities can only increase, and the chances multiply of a grievous miscalculation on the part of either key producers or consumers. It is this deteriorating position of increasing vulnerability to which President Carter has attempted to alert the nation.

The Further Prospect:
Energy Beyond the Twentieth Century

It is difficult to speak of the distant future. Not only are technical and quantitative factors only vaguely perceptible, but, more importantly, the details of the energy situation in the twenty-first century depend heavily on the decisions nations take today and in the near future—or do not take. In addition, we do not know how society will look so far into the future, and surely the nature and structure of the society will have a bearing on energy requirements.

With these provisos in mind, it is possible to suggest that in the years following the turn of the century, the geopolitics of energy may be far less important than it is today. Toward the very end of the current century, electricity will provide an ever-increasing share of energy supply. Nuclear energy and breeder reactors will supply a larger share of electrical generating capacity. Uranium scarcity may be as acute as the current oil situation, but the breeder reactor may be functioning sufficiently to extend the life of uranium resources. Oil usage may be more restricted to its critical uses—transportation and petrochemicals. Contributions from oil shale, tar sands, and solar energy will be more significant. However, it will be a decade or more into the twenty-first century before energy independence becomes possible for a large number of countries.

Nuclear fusion would, of course, be a totally domestic energy source, if it could be operationalized and commercialized. Solar energy also holds out the promise of energy independence, as does—for the United States at least—the maximum utilization of coal. Fusion and solar energy would largely free nations from the constraints imposed by the geopolitics of energy, but not much is expected of them before the first or second decade of the twenty-first century. Could they have the desired effect before the middle of the century?

The challenge posed by the geopolitics of energy is how the world will meet its energy requirements for the remainder of the century, particularly for the next ten, fifteen, or more years, in which oil will remain dominant and in which its location will be sharply restricted to one geographic region. Later, oil will be important as a feedstock and in uses in which alternatives simply are not available; less oil will be used for fuel than for other purposes. The question is how do we survive in the intervening years? How well will the industrial, energy-consuming nations cope with the competition among themselves over access to energy resources? How well will relationships evolve with the energy raw material producers?

Part 2

Policy Options

9
Considerations Affecting
U.S. Energy Policy

The events of 1973 demonstrated that U.S. security and well-being were dependent on a continuous flow of imported oil in adequate quantities, the control of which had passed to the oil-producing governments. Economic well-being and industrial production, domestic political stability and military capability are interrelated, so that the shift in control of oil supplies to the oil-producing states implied that the United States had become vulnerable to the actions of certain countries to an unprecedented extent. Hence, since 1973, U.S. freedom of maneuver in international affairs has been, and is likely to continue to be, unduly constrained by our dependence on oil imports and the necessity to consider the interests and demands of the oil-producing states. While access to overseas energy is only one of our key interests, it cannot be given a low priority; it must stand among our most important. How well we secure our energy interests will help determine our success in meeting other goals.

Without minimizing the essential importance of oil, a nation's freedom of maneuver in international affairs is limited by domestic considerations. This is as true of energy as with any overseas interests, and it becomes increasingly true as the implications of worldwide economic interdependence become clearer. Nevertheless, while oil has peculiar characteristics (e.g., its vital importance to industrial economies and the unavailability of immediate substitutes), the United States does have power to affect developments. The United States has considerable market power by dint of its large size and wealth and economic relations (which gives the United States veto power at least). The U.S. economy's resource position is more favorable than that of most industrial states. Finally, the United States retains a military capability that some oil producers see as a guarantee of their own security; or this might be an asset that could be employed in last resort to obtain supply. No other oil-importing country has this capability.

The post-1973 world offers both constraints and possibilities for international action; our dependence on oil creates problems as well as opportunities.

The need to insulate the domestic economy from forces beyond U.S. control, the need to free military operations from the potential constraints imposed by foreign powers, and the desire to free foreign policy to a greater extent from the constraints of domestic considerations (which should follow naturally from success in the first two areas) constitute a general U.S. objective, which may be described as the need to limit U.S. economic, political, and strategic vulnerability resulting from dependence on energy imports. Yet U.S. energy interests and initiatives cannot—or must not—be considered in isolation, for policies and action undertaken by Washington deeply involve the interests and vulnerabilities of allies. To the extent that allies in Western Europe and Japan remain vulnerable to the actions of energy exporters—as they will for the foreseeable future—efforts to limit U.S. vulnerability must consider the effects upon Western Europe and Japan. The reverse is equally true: pressure can be brought to bear on the United States through the continuing and inescapable import dependence of its allies. At the very least, the U.S. economy could not remain unscathed by adverse economic and political developments in Western Europe and Japan resulting from inadequate or sporadic oil supplies.

A critical case in point is the competition anticipated among governments for Middle East oil, competition that is bound to intensify over the next 10-15 years.

The United States will be concerned with the divisive

effects of competition for oil within the Western and Japanese alliances, but it will also be concerned with its own competitive economic position vis-à-vis its allies and with how this competition may affect its position vis-à-vis the communist nations. Moreover, it will have to be concerned with the competitive position of the Western alliance generally vis-à-vis the Warsaw Pact.

Since this competition will develop owing to the paramount producing role of the Gulf states, and since oil in world trade will increasingly originate from that region, the United States must pay a great deal of attention to such factors as the stability or direction of political and economic change in the region, the limiting of or countering the extension of Soviet influence in the area, the relative positions of Saudi Arabia and Iran, and the security of the Straits of Hormuz and the Red Sea. Within the Gulf, Saudi Arabia will be pivotal and thus an area of profound interest and concern to all energy-deficient states. The special relationships the United States has created with Saudi Arabia and Iran with regard to military and economic assistance are naturally viewed as embracing oil as well. The present paramount role of the United States in the Gulf also gives it a special responsibility—in the eyes of other importers. In this regard, traditional ways of linking producers of raw materials to the interests of consumers—expressed in terms of an imperial system—give way to still ill-defined modes more appropriate to a different period.

The United States must be concerned with these new avenues and mechanisms for securing access. These may evolve, with or without U.S. participation, and they will create precedents for energy in world trade. More broadly, the issue of access to raw materials, including oil, is linked to developments in "North-South" relations generally.

Energy import dependence is not limited to oil, but includes West European and Japanese potential dependence on Middle Eastern or Soviet gas exports, and a possible U.S., West European, and Japanese dependence on imports of natural or enriched uranium. In the area of natural gas, the

United States must be concerned lest its own increased imports of LNG, as well as Europe's and Japan's, increase
energy dependence on the Soviet Union or multiply dependence on the Middle East. The nuclear fuel and technology
dependence of most developing and industrial economies
rests on a handful of states, with the possibility that access to
the basic raw uranium resource may again raise issues comparable to those posed by oil.

Part 1 of this study postulates Free World dependence on
energy imports at least through the end of the century. The
central issue is how to navigate through the interim, during
which dependence on imported energy sources persists and
before the constraints imposed by the geopolitics of present
forms of energy may be lifted. Domestic politics and foreign
policy objectives beyond access to energy will sharply affect
our passage. In the following chapters we will look more
closely at the contemporary issues that will be of concern.

10
The Nature of the Threat: Emergency Situations

Before an energy policy dealing with long-term energy considerations can be formulated, it is necessary to make sure that there will, in fact, be a long term. Only after it has the capacity to overcome short-term, occasional shocks can a nation undertake the careful, time-consuming, complex analysis needed to elaborate long-term objectives and policies. The so-called emergency situations must be attended to first, and in order to accomplish this, the nature, extent, and scope of the potential threats must be defined. Having done this, policy can then be devised to serve two interrelated goals. First, policy can be designed to deter a threat from arising. Second, it can be designed to negate the effects of a threat that has already been carried out. Clearly, the second is an important element of the first, but it does not exhaust the possibilities for deterrence.

Embargo

From 1973, the most obvious threat has been a deliberate embargo on oil shipments to the industrialized countries, either collectively or selectively, by the oil producers acting in pursuit of some objective of their own. Given the disparate nature and interests of the OPEC countries, we doubt they share sufficiently a common political purpose that all members would be willing to support to the extent of halting all oil exports to the industrialized world. The collapse of "North-South" discussions might be such an issue, but a more likely response to Western intransigence appears to involve higher prices or some cutback in production or both rather than a total embargo. In fact, a general embargo in support of LDC objectives has never been proposed, would probably involve much higher costs than OPEC is prepared to pay for LDC diplomatic support, and would depend on a greater, more united commitment to general LDC goals than probably exists within OPEC. However, higher oil prices and curtailed production are quite serious enough.

In addition, an OPEC-wide embargo clearly involves heavier costs for some OPEC members than for others. Given their internal political, economic, and social systems, some OPEC members are better able to sustain periods of lost revenue than others. Saudi Arabia, Kuwait, the United Arab Emirates, and Qatar—all countries whose income exceeds their present revenue needs—may be able and willing to sustain an embargo with little negative domestic repercussion. Iran, Venezuela, Indonesia, Nigeria, Algeria, and Iraq are countries with large populations and ambitious development plans, which make oil revenues essential. For the next several decades, the likelihood of this latter group's participation in an embargo for any but their own pressing interests is debatable. The position of other OPEC members is still more uncertain, but they do not produce enough to be important: Ecuador, Gabon, and even Libya.

The narrower, all-Arab OAPEC, however, includes those countries that are perhaps capable of sustaining a loss in revenue and that share a sense of common political purpose derived from the Arab-Israeli conflict, or from pan-Arab nationalism. In 1973, OAPEC instituted the embargo and the associated production cutbacks. For domestic political reasons, an embargo in support of an Arab cause might be attractive to these countries. (Even here, however, the extent of unity can be questioned; Iraq, an anti-Israeli militant, did not participate fully in the 1973 embargo.) In an embargo of short duration, the wealthier OAPEC states could probably subsidize the poorer oil producers for their participation if it were necessary to include them. However, as OAPEC will be increasingly the most important source of oil supplies (and Saudi Arabia the most important single source), even an embargo that excluded Nigeria, Gabon, Venezuela, Ecuador, Indonesia, and Iran might produce a situation in which oil supplies from these nonparticipatory states could not compensate for oil lost in the action.

"Conservation"

Another category of possible producer state action could have effects similar to those of an embargo. Different objectives of different states might coalesce in conservation measures. A major producer, or several of them, could institute "conservation" measures and thereby reduce supply, perhaps in graduated steps.

Conservation measures may be entirely legitimate, i.e., based on objective facts regarding field characteristics or surplus revenue accumulation. Such measures may also be used to tighten the oil market to sustain prices. Some producers may simply want to dampen the rise in excess revenues. For others, "conservation" measures might be used to put more political pressure on consumers. Elements of all four might serve among disparate producers as one or more reasons to institute measures to restrict production. Since petroleum experts generally warn of a tightening supply situation—toward the mid-1980s—"conservation" policies could quickly take effect and be less incendiary in terms of importing states' reactions than an embargo would be; this action by producers would also be more difficult for consumers to counter through their International Energy Agency, which is designed mainly to cope with embargolike events.

The recent fires that shut down some North Sea and some Saudi production are also indicative of emergency situations that may arise in the future.

Conflicts in the Middle East

Beyond the threat of a deliberate oil embargo of varying proportions and beyond the threat posed by conservation policies, an additional threat to secure and adequate oil supplies might originate in conflicts involving the Middle East or within the Gulf, conflicts in which a sudden oil shutdown by one nation or another could be expected. A long-expected variation is one in which animosities among the Arab states, or between some of them and the Iranians (complicated by U.S. relations with both Iran and Saudi Arabia), could shut off very large volumes of oil, even if only sporadically. The Israeli "issue" is, of course, of great, continuing concern, but it is not the only dispute that could inflame the area.

Today, the Israeli "issue" looms as the most likely confrontation that could again plunge the Middle East into war. In the event of war or should Saudi and Egyptian efforts to set the stage for some form of political settlement soon come to naught, it is only prudent to take the Saudis at their word and to expect that a certain consequence would be employment of the oil weapon on an unprecedented scale.

Saudi actions in this regard would probably be supported by Iraq and Kuwait, possibly Libya, and certainly by the Emirates. Syria's contribution, as far as oil is concerned, would be negligible. Iran might again continue supplying, but nothing is certain in Middle East political alignments, and a very great deal would depend on the circumstances at the time the oil weapon was taken up. In any event, given the growing importance of Saudi oil, we will reach a point where Iranian willingness to supply the industrialized countries cannot compensate for lost Saudi exports.

In more recent years, there has been speculation that in retaliation—or even in a preemptive strike—the Israelis might attempt to destroy the oil facilities of the Gulf. A preemptive strike might halt or slow the financing of arms supplies to the confrontation states by the major oil producers. Assuming the Arabs continue to regard the Israelis as having practical mastery of the air, the possibility of such a step might be the most potent deterrent to Saudi or Iraqi actions against Israel or to Saudi support for military actions by the confrontation states (Egypt, Syria, and Jordan). It would be an additional factor pressing the Arabs toward a political settlement now. An Israeli attack on the oil facilities would engulf the Middle East in a maelstrom with such consequences for all that a decision to "take out" the bulk of Gulf production would bring all parties to Armegeddon.

The pivotal actor in all these possibilities is Saudi Arabia. The Saudis alone might not be able completely to offset the effects of any of these developments, but their participation could guarantee success. Their nonparticipation in a shutdown might be sufficient to avoid catastrophe in the industrialized countries. Under any circumstances, the Saudis would be under intense political pressure from all sides. The political future of this sparsely populated country, which has a resource vital to others, is accumulating substantial financial resources, and is located in a highly unstable region, will continue to be of utmost concern to all energy import-dependent countries. The other potential actor is, of course, Iran, whose ambitions, military power, and insatiable need for revenue could find access to another's oil irresistible.

Responding to Emergency Situations

International energy cooperation to date has been geared largely to meet emergency situations posed by producers against consumers. It has centered on efforts to: (1) demonstrate the seriousness with which threats are viewed by the industrialized countries; (2) deter the implementation of threats; (3) negate the effects if the threat becomes action; and (4) avoid divisive competition among consumers in the event of shortage.

The encouragement of energy cooperation among the industrialized countries has been very much a U.S. initiative.

It has met with less than full enthusiasm in either producing or consuming states. Most producers assert that cooperation among the industrialized nations encourages a policy of producer-consumer confrontation, which, they have argued, benefits no one. Their real argument is that the International Energy Agency (IEA) can vitiate the effectiveness of an embargo only partly.

Consuming states that are more dependent on imported oil than the United States is are seriously ambivalent about the cooperative effort. International energy cooperation is beneficial to the extent that it avoids higher costs and prevents threats from being carried out. At the same time, as producers denounce "confrontation," heavily dependent consumers become increasingly nervous. Moreover, to the degree that consumer states judge that their link with the United States multiplies their risks, international energy cooperation in the framework of the International Energy Agency could become too expensive.

Still there is a clear necessity for having on hand responses adequate to emergency situations. We have the elements of the requisite program in the International Energy Agency.[1]

The near-automaticity of the sharing agreement must obscure neither the differing risks members believe they run from their association in the IEA nor the low probability that the overly complex and nearly incomprehensible formula for sharing can be implemented. Rather, the reasonable expectation is that the international oil companies will again be asked to share available oil as equitably as possible, and to do so under the aegis of governments.

Assuming that another embargo is initiated, which nations are likely to be the targets? Increased U.S. dependence on Middle East oil could make the United States a direct target, particularly if the provocation for the embargo were again the Arab-Israeli conflict. However, we now realize that the targets could include Western Europe and Japan with the intention of bringing indirect pressure to bear on the United States. The fact that Europe and Japan are now linked to the

United States in the emergency sharing formula almost guarantees their involvement in an embargo. Even if the oil producers permit oil exports to Europe and Japan on levels not exceeding preembargo shipments, the need to share would reduce oil supplies available for domestic use. The linkage can be seen to multiply the risks incurred by Europe and Japan. The very possibility that Europe and Japan may have to bear the brunt of Arab displeasure with U.S. Middle East policy could have intensely divisive repercussions within the Western alliance, which would become evident at the moment of crisis unless the United States has foreseen the possibility and shapes its response intelligently.

The International Energy Agency's emergency sharing program could well negate the effects of a threat made good. If the United States is the target, however, the Europeans and Japanese are committed to share available supplies with the embargoed country, an action certain to incur the anger of the oil exporters. If Western Europe and Japan are the targets, however, the burdens will be on those countries to share oil—more so at least than the United States can contribute to their supply and try to offset the embargo. It seems likely that the emergency sharing formula serves as something of a deterrent; if deterrence fails, the IEA formula may well be adequate, although some sharing could occur through ad hoc measures. The actual implementation of the IEA provisions will depend on the circumstances.

Since the IEA arrangement does introduce an element of uncertainty into government calculations, it cannot now be abandoned without suggesting to the producing countries that the industrialized countries are as divided and weak as the former believe them to be. Still, the United States must anticipate that the Europeans and Japanese will be extremely reluctant to invoke the emergency sharing formula. It should even be quite willing not to ask for invocation in the first place. In any case, the United States may well have to be generous in its understanding of what consequences an embargo will have in the more oil import–dependent nations of

Europe and Japan.

The essential operational characteristic of an embargo is that it accomplish its objectives in a short period of time. The longer the embargo, the more drastic the economic consequences to the industrialized countries (this will have negative economic repercussions in the oil-producing countries and other LDCs as well) and the greater the chances that, in desperation, military action will be taken against a producing state. This is more fully understood today than it may have been in the emotional atmosphere of 1973-1974; will it be remembered in another crisis?

Stockpiling, or strategic reserves, is now accepted not merely as a way to tide a country over a period of embargo but also as a way to meet the possibilities that conservation policies or accidents will limit available oil or that upheavals in the Middle East will disrupt supplies. The indeterminate effects of stockpiling must be recognized; stockpiles and reserves provide some protection against supply disruption, but they may also provoke a longer, more geographically widespread, and quantitatively larger cut in oil supply. That is, the producers may have a sense of greater commitment once they have decided upon and carried out an embargo. Moreover, stockpiling efforts imply that the producers must be prepared to sustain greater and longer-lived financial and other costs. Thus a national strategic reserve provides an essential tool not for embargoes alone but for supply situations short of an embargo. It serves both a deterrent and an "after-deterrence-fails" function. But as stocks are drawn down, the pressures on producers and consumers for some form of political settlement increases: the consumer nations approach an energy "deadline" after which the threat of economic collapse makes action seem unavoidable. The producer nations should be aware of the risk they run.

In addition, a very large consumer-importer, possessed of a sizable strategic reserve, could apply pressure against a major supplier by suspending imports. Depending upon the world oil market at the time, such an initiative could severely damage an oil exporter.

Petroleum Stocks*

In normal circumstances, commercial stocks, or inventories, are a vital part of the oil industry's distribution system. When the normal flow of oil is disrupted, stored oil can "fill the gap" for a limited period of time. The 1973-1974 Arab embargo and subsequent government activities have made this second role of oil stocks very important indeed. At present, and through the 1980s, oil stored in each importing nation will be a key factor in limiting vulnerability to accidental or deliberate disruptions in world oil supplies.

Stock Information

In discussing the protection that current and prospective levels of stocks give the major importing countries, it is necessary clearly to distinguish among military reserves, commercial stocks, and strategic, or emergency reserves. Military reserves are generally kept separate from civil supplies and are for military requirements.

Commercial stocks, or inventory, are held by oil companies and sometimes also by major consumers, such as utilities. An oil company will hold stocks of crude oil and product emerging from its refinery. Oil in pipelines, tankers, barges, trucks, and rail tank cars are all part of the commercial stocks that oil companies report to national governments and that are included in the stock figures published in statistical surveys. The statistics, when reported weekly or monthly, will show large variations in particular product stocks on account of seasonal fluctuations in demand.

These commercial realities are reflected in stock statistics, and they make it extremely difficult to distinguish in the data between commercial stocks and emergency reserves. This is particularly the case in countries where governments

---

*The following section was very largely contributed by Edward N. Krapels, at the time a Research Fellow of the London-based International Institute for Strategic Studies, and an associate of the authors.

do not require that industry maintain an emergency reserve, or where they have only recently done so, as in West Germany. (The United States has only recently launched a program to create an emergency reserve.) No one can yet estimate accurately how much stored oil in the industrial states could be used in an emergency before the reduction in stock levels begins to cause spot shortages and even runouts.

The most thorough assessment in this area has been conducted by the U.S. National Petroleum Council (NPC) (September 1974). For the American oil market as of September 1973, 170 million barrels of primary crude and 140 million barrels of primary product stocks were simply "unavailable"; that is, when inventories are drawn down to that number of barrels, the "United States logistics system simply would not operate." This is oil held in pipelines, tank bottoms, and in retail delivery vehicles. An additional 70 million barrels of primary crude stocks and 360 million barrels of primary product stocks are needed by the industry to maintain "continuity" of operations. Thus, the "minimum operating level" of stocks, which is the sum of unavailable oil and stocks needed to maintain the continuity of operations, amounted to 240 million barrels of crude and 500 million barrels of product.

In September 1973, by these definitions, 99 percent of reported crude stocks and 97 percent of reported product stocks were needed to support the distribution processes of the industry. According to the NPC, none of these can be considered true emergency stocks or reserves.

The NPC's high estimate of minimum operating levels implies that the American oil industry is very efficient indeed or that the NPC is excessively cautious. In fact, both are true. There is strong evidence to support the claim that the nation's oil distribution system is efficient. The NPC notes that although demand for oil products increased by 28 percent from 1968 through 1973, and refining capacity expanded as well, total inventories did not rise. Furthermore, stocks are a significant drain on an oil company's cash flow, especially at today's oil and money prices, and in the absence of government requirements, the oil companies have no incentive to hold more stock than is absolutely necessary. In a subsequent report (August 1975), the NPC made a somewhat more generous estimate of what it called the "United States emergency reserve base": as of December 1974, about 140 million barrels, or roughly twenty-three days of American oil imports at that time.

Europe and Japan

There is no equivalent to the NPC assessment for Europe and Japan. In fact, statistics on the stock levels of these countries are generally not reported to the public. A listing of "stock" levels for all OECD countries can be found in *International Oil Developments: Statistical Survey,* published by the Central Intelligence Agency. A 1975 listing of stocks is shown below, the situation has not changed in any important respect.

Table 25. Oil Stocks in Europe and Japan

Country	Stocks (thousands of barrels)	Days consumption[1]	Notes
Japan	324,000	68	Average monthend for 1975.
Britain	140,000	74	June through December 1975 only.
France	214,000	97	Average monthend for 1975.
Germany	169,000	63	Excluding January and April 1975.
Italy	157,000	83	February, June, July, August 1975 only.

[1] Utilizing final inland consumption data found in "BP Statistical Review of the World Oil Industry."

The CIA does not divide stocks into crude and product categories. Needless to say, it is difficult to conduct even a rough country-by-country assessment of the level of commercial stocks versus emergency reserves from such data.

Nevertheless, it is certain that a large part of European and Japanese stocks falls into the "unavailable" and "minimum operating level" categories to which the NPC refers. But since European markets are more compact, have much less pipeline mileage, and more central refinery operations, and since the oil companies are required by law to maintain a certain level of inventories in some countries, it is likely that the share of inventories available for emergencies is larger than in the United States.

EEC country governments have been under pressure from the commission to require that the equivalent of ninety days of consumption be held in stock at all times. It has recently been reported that the commission is taking Belgium, West Germany, Ireland, Italy, and the Netherlands to the European Court for failure to establish this legislation. Yet the Community as a whole reports 108 days of consumption in stock. The Netherlands was reported to have 236 days (probably includes stocks held by export refineries), Denmark 141, France 118, Belgium 116, Germany 93, Britain 89, Luxembourg 79, and Ireland 76.

These estimates are meaningless without some knowledge of the consumption statistics against which they are drawn. The commission, however, is not releasing more specific information. Energy specialists in the commission regard the analysis of the real level of emergency reserves as important EEC and national government intelligence.

The OECD and its energy branch, the International Energy Agency, similarly have not made public the members' stock positions. (The OECD is now in the process of issuing non-restricted data on crude and product stocks to member governments.)

We continue to be limited, therefore, in our ability to assess the extent to which Europe and Japan are protected from oil import disruptions. Oil industry sources have provided estimates that European countries, on average, must maintain thirty days of crude and thirty days of product, measured against daily consumption, to maintain the continuity of their oil distribution systems. Using this estimate, we can infer that in selected European countries, the emergency reserve base, as measured against the data given earlier on country stocks and consumption, is fourteen days for Britain, twenty-seven days for France, three for Germany, and twenty-three for Italy. Measured in terms of the number of days of imports, the situation in France and Italy is roughly comparable to the United States, that is, about 20-30 days of imports.

Stocks, Imports, and the International Energy Program (IEP)

The number of days of imports a country has in emergency reserves has become a more confusing question since the formation of the International Energy Agency. Initially, the negotiators sought to reach an agreement on the level of international emergency reserves; these reserves play a crucial role in the manner in which the IEP's designers hope to manage a future crisis. The negotiators then settled for an interim agreement that adopted the OECD definition of stock levels (never intended for anything but often politically inspired reporting purposes), deducted the oil held in pipelines plus an additional and arbitrary 10 percent as unavailable, and labeled the rest "emergency reserves."

Although the IEA has not released estimates of the emergency reserves each member has under this definition, it is possible in the case of the United States to work back from the definition to arrive at the IEA's estimate of the emergency reserves. The NPC did this, and the resulting figure is about 950 million barrels. The difference between this and the NPC's own estimate of 140 million barrels is striking. The NPC commented that the IEA estimate "does not recognize the need to keep the oil logistic system operating efficiently." A similar judgment could probably be made with respect to the other IEA members, although it is not possible to be as precise.

We can, however, give a rough indication of how long stocks could protect the IEA members against several kinds of disruptions.

In spite of the inadequacy of the data, these estimates are of interest because they show how wrong the data would

Table 26. Free World Oil Reserves (in days)

Selected countries	Normal consumption 1975 (thousands of barrels per day)	Estimated emergency reserves (millions of barrels) [1]	Number of days emergency reserves would last if drawdown equals following percent of normal consumption		
			10	20	30
United States	15, 845	140	87	44	22
Japan	4, 905	39	80	40	20
Britain	1, 875	27	142	71	36
Germany	2, 665	8	30	15	8
France	2, 240	81	370	135	68
Italy	1, 925	43	226	113	57

[1] Assuming the "minimum operating level" is equivalent to 60 days of normal consumption, excepting the United States, for whom the NPC estimate is used.

have to be before the estimates would show that countries could rely on current stocks for a significant period of time.

Prospects

While conventional stocks have not been appreciably built up since 1974, great plans are being considered by some governments. The United States appears committed to creating a separate "strategic petroleum reserve" of 1,000 million barrels by 1985. The government of Japan recently declared its intent not only to achieve ninety days of consumption in conventional storage, but also to create an additional ninety-day reserve under government control. The European countries, except France, have more modest goals. The IEA may be influential in prompting further increases, and individual countries have embarked on programs of their own. On record are Germany, with the modest objective of increasing federal government stocks from the current 20 million barrels to about 60 million barrels, and France, which is a "closet" member of the IEA and whose government has decreed that all crude importers and refiners, not just distributors of finished products, are now required to maintain forty-five days of stock.

The American and Japanese reserves will make a substantial difference in the length of time a given oil import loss can be maintained. Roughly speaking, if the United States achieves a 500 million barrel reserve, it can quadruple the number of days the shortages can be sustained. However, if the United States' imports and dependence on the Middle East continue significantly higher than they were in 1975, the length of time is shortened. The same point can be made for Japan. Europe's vulnerability will not be significantly reduced with the current stock plans.

Finally, an assessment of the role of stocks must consider the adversaries against whose action they are to be used. The Arab states currently provide about 60 percent of the imports of the United States, Japan, and Europe, or over 21 million barrels per day. Should they decide to curtail production, they must take account of the increase (if any) in emergency reserves in their assessment of how to put pressure on the importers. If they want to confront the importers with an irreplaceable loss of oil supplies quickly, they must curtail production significantly to force the importers to draw down their stocks. It is estimated that if the Arab exporters cut back by 50 percent, resulting in an import loss of roughly 25 percent and a supply loss of roughly 18 percent, stocks would be able to maintain a consumption level of 90 percent for only a few months in many countries. Some will face the loss more quickly than others, which further complicates the attempts of the IEA to create a united consumers' response.

In sum, current stocks can "buy time," but not much, and they do not provide absolute security against determined adversaries. Even when the level of emergency reserves for the United States reaches 500 million barrels, our own higher import dependence and the continued vulnerability of the European countries will still be cause for grave concern. Meanwhile, the incongruity of industrial states not yet being able or willing to confront their energy vulnerability in terms of their commercial stocks versus true emergency reserves—a situation that has now been with us for several decades—is inexplicable.

Conclusion

Energy cooperation among the industrialized states involves relations between consumer states and between them,
severally and collectively, with producers. With regard to the
importing countries, cutthroat competition among the consumers may have been forestalled by agreements within the
International Energy Agency. Nothing in the emergency plan
precludes closer relationships between individual consumers
and producers, but it does make the search for such relationships less frantic.

Various responses to possible emergency situations are
necessary, but they are only palliatives. Real amelioration—
or even the avoidance of emergency situations—depends on
other actions, actions taken with the longer term in view.

11
The Nature of the Threat:
Longer-Term Supply
Security Measures

Emergency stocks, conservation and rationing, and stand-by agreements to share available oil may suffice to permit the energy-deficient industrial nations to withstand embargoes and sudden supply disruptions. They are clearly not answers to the problems created by the geopolitics of energy: the location of energy resources in countries other than the major consumers and the need of the industrialized nations to obtain reliable access to them. There is, in addition, another set of factors that are peripheral to long-term solutions but that nevertheless reflect concerns arising from the geopolitics of energy. Secure energy supply will have to include policies relating to critical segments of supply logistics.

Three aspects of this longer-term situation have been identified in Part 1. They include the need for a careful and continuing watch over refining capacity, tanker control, and the security of "superport" facilities and offshore terminals.

Refining

With regard to refining, there is presently a very considerable spare (unused) capacity throughout the world. Generally, the United States shares in this situation if the Caribbean refineries that have traditionally supplied the U.S. market are included as if they were "U.S." The prospect, however, is that with anticipated economic recovery, the idle capacity that plagues the industry today could disappear over the next five years. In the case of the United States, its "idle capacity" (based upon these Caribbean refineries) will fade considerably sooner, and the United States may be increasing its product imports. When this occurs, and if there is no carefully timed introduction of new refineries, U.S. vulnerability to supply cutbacks from abroad could be considerably increased. Without those Caribbean refineries, domestic refining in the United States itself is now at capacity.

As indicated earlier, crude imports have a flexibility in sources that permits—within limits—the substitution of one geographic source for another; product imports cannot be replaced so readily, especially when there is no idle domestic refining capacity. Currently, the spare capacity to meet U.S. need beyond the Caribbean facilities is in Europe.

This product vulnerability (generalized to include Europe and Japan) could result from a situation in which refineries in the consuming countries are rendered "uneconomic"; the producing countries may expand their refining capacity and insist on supplying more product and less crude. This has been a long-standing OPEC objective. While there is no early prospect of such a development, OPEC investments "downstream" bear watching.

Government and industry must anticipate such developments and take actions to assure a U.S. domestic capability (defined as including Caribbean refineries) to meet all our product needs. Some form of public subsidization of refining capacity may become necessary, much as stocks of key materials are being created.

Tankers and Port Facilities

With regard to tankers, until the United States has deepened its harbors and completed its superport facilities offshore, it will not be able to take full advantage of the economies of scale provided by the giant tankers (VLCCs). The largest tanker that can now enter U.S. harbors is still about

85,000 DWT (although larger vessels are often lightened). Tankers of this size no longer dominate DWT in the world oil trade; in fact, they have become anachronisms as far as major supply to a giant industrial nation is concerned. The number of tankers in this category is diminishing, and there are few orders for replacement. Yet it is precisely on tankers of this size that U.S. supply will be forced to rely for at least the next ten years.

Maintaining an adequate fleet of such vessels is not only a commercial requirement but a matter of national security as well.

In addition, governments may have to encourage greater investment in product tankers, since more and more of the world's oil trade will take this form. The possibilities for further conversion of some surplus crude oil tanker capacity should also be investigated.

The problem of defending critical offshore terminals, the concentration points for immense volumes of oil, can only be noted. As an example of the importance of these facilities, the single Shetland Islands installation, Sullom Voe, which is to receive Shell/Esso and BP North Sea production, could become the reception facility for over 50 percent of Britain's oil. Similarly, terminals in the producing countries that process large volumes of oil for export, such as Saudi Arabia's Ras Tanura and Iran's Kharg Island, must be kept secure. For the United States, if LOOP and SEADOCK deepwater ports were built, perhaps as much as 75 percent of U.S. imports would be received at these two stations only. The security implications are obvious.

12
Limiting Vulnerability: The Domestic Component

Means to limit U.S. vulnerability to sudden, short-term supply disruptions and threats to the logistics system are clearly essential. In an important sense, however, these are exceptional occurrences, responses with which to buy time, and thus permit more fundamental actions to limit our continuing vulnerability. By concentrating on the United States, we do not imply that many of our options are irrelevant to our allies; indeed, with some variations to reflect individual cases, comparable opportunities exist for others, acting singly in some options, and acting together when useful to do so.

Comprehensive Energy Policy

Obvious as it may be, the first prerequisite for limiting U.S. vulnerability is that the United States must adopt a compehensive National Energy Policy and Plan. The statement of goals and policy directions by President Carter is only the beginning. Until Congress passes legislation to implement these goals, we cannot say that the inadequate energy record of the past is behind us.

The absence of an overall energy policy doubtlessly projects confusion and continued vulnerability, and it could suggest to producing governments that U.S. military action may be the only possible response to an action of theirs that runs afoul of U.S. interests. In this instance, it has been held that deliberate vulnerability can serve as a kind of deterrent moderating the behavior of producing countries. The element of uncertainty is allegedly made to work for the United States. However, this argument is more an excuse for failure to formulate energy policy and to make difficult, but necessary, commitments.

A comprehensive energy policy, on the other hand, demonstrates the seriousness with which the United States views energy matters and its determination to be freed from the constraints imposed by the geopolitics of raw materials. Internationally, a comprehensive energy policy reveals the seriousness of America's commitment to adequate and secure energy supplies: "energy" unambiguously and profoundly receives U.S. attention and commitment; it is not to be left to the aims of foreign nations blessed by a geographical accident with a disproportionate share of the world's oil resources.

A comprehensive energy policy holds out the possibility that the dominance of the producing states may well be temporary and, it may well induce caution among producers. While not suggesting a dilution of effort, we recognize that if we make the position of the producing states less secure in the long term, some might be induced to exploit their temporary advantage more vigorously. Against such possibilities, the "defensive" measures of strategic reserves, adequate tanker control, and refining capacity make good sense—if done by all.

These requirements would allow the United States, then, to proceed on its energy initiatives. Domestically, they are: (1) conservation of all fuels and their highly selective use, (2) the intensive development of coal, (3) the search for additional indigenous and hemispheric supplies of oil, gas, and uranium ore, (4) the maximization of efforts to bring nuclear energy into our economy on widely accepted terms, and (5) energy research.

There is action on these fronts; the essential point is that actions taken to date have been ad hoc, with little recognition of the need for an energy policy and commitment that acknowledges the interdependence of all parts of the energy picture. Moreover, actions to date, emerging from fierce battles between the administration and Congress and within the administration itself, have done little to reduce the vast areas of uncertainty that continue to inhibit the private sector.

The president's energy statement falls short of a comprehensive national energy program for several reasons: the emphasis on conservation with so little attention to production; the references to coal without consideration of the ancillary areas (amendments to the Clean Air Act, transportation facilities, and others) that must be addressed if coal is to make a greater contribution to the nation's energy supply; and the unmentioned, but significant, fact that even if the Carter package were adopted in full immediately, the nation will continue to require substantial oil imports.

With regard to the longer-term search for additional oil supplies, Part 1 offered a number of observations concerning potential, highly prospective oil regions. The United States can embark on an unprecedented effort to expand supply sources outside the Middle East to three highly prospective areas within this hemisphere, and also to the sub-Arctic and Arctic regions, where control over discovered oil would be within the grasp of the industrial world.

These three areas are Canada, Mexico (and the Caribbean basins), and the Orinoco belt of Venezuela. Success in developing oil resources in these three, along with the strategic crude and product reserve, should give solid assurance that the United States can withstand contrived supply shortages. The political interests of these three are not related to the regional political interests of the Gulf (nor are they likely to be participants in embargo actions taken against the United States over Israel).

Moreover, the United States should consciously reduce imported crude and products whose origins are in the Gulf. While we have pointed out the security implications of the Gulf Straits and the African and South Atlantic routes, we do not ascribe to the view that oil shipments from the Gulf are inherently and continually in danger. The interests of all key producers—except possibly Saudi Arabia and Kuwait—are embedded in supply continuity, but prudence, reinforced by the desirability of keeping the United States flexible, indicates that preferential import selection away from the Gulf is desirable.

Attainment of a billion barrel emergency stock and the ability quickly to introduce a stringent, accelerated conservation effort or rationing or both are still the most important steps the United States can take to limit its immediate vulnerability to supply shortage. A reserve of this magnitude is probably more than ample (except in the aftermath of a general nuclear war).

However, some assert that the creation of a strategic reserve may be all that is necessary to render us all but immune to any plausible turn of events. This is nonsense. The reserve is sufficient only to buy time to permit a state to exercise all its applicable powers to resume the flow of oil. The reserve does not, of course, affect the level of imports, encourage indigenous energy resource development, or meet other requirements. As noted, attention to the need for adequate domestic refining capacity is also required; with a comprehensive energy policy, we would be in a position to examine how to maintain that level of refining capacity needed under U.S. control or influence. Similarly, we could look again at the perennial question of effective U.S. control over crude and product of the required amounts.

In the area of nuclear energy, we still require completion of a massive survey of the United States in order to determine realistically available indigenous sources of uranium ore and thus the level of necessary uranium ore imports. We require a realistic definition of the additional enrichment or other processing facilities adequate, with some spare, to meet national and international needs. None of these surveys yet exists, and none can be obtained easily. Yet without authoritative assessments, nothing intelligent can be accomplished either in the framing or implementation of policy.

Finally, a comprehensive National Energy Policy must begin the process of defining the role of industry and government as we move into an energy era of unprecedented complexity. As in so many respects, our dealings with the really new and important set of relationships, as has been implied throughout this study, have been ad hoc and inadequate. They satisfy no one: government, industry, or the public. Yet this evolving relationship must be clarified if we are to engage fully the great economic, technical, and managerial assets of the private sector.

Role of Government

Over the past several years, there has been an unending stream of recommendations as to what must be done, by whom, and how quickly. Many of these deal with real and vital interests and objectives; most see an unprecedented role for government in assuring that the nation's needs are met.

Few of these proposals emphasize the complex, interrelated, and interdependent actions that must occur in a timely manner if present, conventional, and future indigenous energy sources are to be available. The cases of coal and nuclear energy are excellent examples of the need for a government role to facilitate those undertakings whose scope puts them beyond the means of private companies. Moreover, until the government takes actions in these areas, private activity will be inhibited by uncertainty.

Furthermore, the U.S. government must be able to assure performance, i.e., to guarantee that an oil company that is also in coal cannot negate a National Energy Policy objective that seeks to limit the use of oil and expand the use of coal. This calls for an unprecedented kind of peacetime relationship, which must be thought through with exceptional care; it does illuminate the hard fact that the provision of energy is no longer only a commercial undertaking.

It will seem anticlimactic, but emphasis must be placed on the absolute, fundamental requirement of a proper and effective national government supervision of energy, even involvement in its supply when necessary. No matter how pressing energy priorities may be, no matter what level of investments and technology are available, the goal of significantly reducing our present and foreseen vulnerability to shortfalls in supply cannot possibly be reached unless the federal government has a sensible, disciplined, and competent grasp of the energy problem. Without government supervision, the most that can be done is to pursue on an ad hoc basis certain aspects of our energy needs in an haphazard, contradictory, and inadequate manner; "energy" could then truly become the nation's "Achilles' heel."

Public support for commitments will be forthcoming, and the private, commercial, and research interests of our country can respond if there are intelligent directions, priorities, and incentives and if national policies are durable. If the United States sets a strong energy tone, its very accomplishment will moderate the attitudes and actions of producer (and consumer) nations who might otherwise doubt our national resolve.

13
Limiting Vulnerability: International Implications and Options

Between Autarky and Vulnerability

In adopting a comprehensive energy policy, the international repercussions must be considered. At one extreme, policy tends toward autarky; at the other, it tends toward deepening dependence. Both these choices involve important international ramifications.

Autarky may be neither possible domestically (in terms of timing, or at reasonable cost, i.e., costs that allow the simultaneous pursuit of other societal goals) nor desirable internationally. If the United States deliberately minimizes its direct involvement in international energy, it may appear to lack interest in and commitment to developments that may be of vital interest to allies (and to the USSR).

However, commitment can be demonstrated in many ways, and it is unlikely that anyone would doubt U.S. concern in an area so vital to its major allies. The point is not to discredit U.S. movement toward greater energy self-sufficiency but rather to note that every policy involves costs and implications, which must be considered if they are to be offset or amplified, as the case may be.

At the same time, greater U.S. energy self-sufficiency could free considerable quantities of oil for use by consumers with fewer energy options than the United States. It was largely the U.S. entry into the international oil market in a big way after 1970 that changed a buyers' market into a sellers' market and that was a very major contribution to OPEC's success.

At the other extreme, the United States could attempt to multiply its own dependence on particular producing states as clear evidence of an enduring interest in them as suppliers. In this way, all the world would be warned that threats to (or from) them would so seriously and directly affect the United States that they would be considered provocative and likely to elicit a serious American response—including military action. Between these two extremes, the United States must find a way to demonstrate its sustained vital interest in the producing states without backing itself into a corner from which only a military escape is possible. On balance, the risks associated with greater U.S. energy self-sufficiency appear to be less momentous and more easily negated than the risks associated with deliberate U.S. energy vulnerability.

Energy, the Western Alliance, and Japan

A comprehensive U.S. energy policy will have important implications for Western Europe and Japan. In fact, the continued lack of a comprehensive energy policy is at least one major reason for the inability of other countries to develop their own energy policies; the uncertainty derived from both the United States and the oil producers makes planning nearly impossible. To some, divisiveness within the Western alliance looks more and more like simple European pique at its dependence on the United States, but the strains within the alliance are nonetheless real.

If the United States obtains a disciplined and intelligent National Energy Policy and Plan, this uneasy situation might be somewhat ameliorated: the various elements of U.S. energy policy could at least be submitted for review to allies, perhaps through the IEA, however tricky it might be for domestic political reasons. (The OECD Energy Committe attempted something like this "show and tell"; it was a failure.) We are aware of current plans to discuss within the IEA broad energy policy goals with provision for periodic reporting by member states on their progress in reducing their dependence on Middle East oil. The initiative is a move in the right direction and should be broadened to include objectives beyond reduced dependence on Middle East oil. If done with frankness and with the recognition that allies have something useful to say, it would seem, on balance, that intermational discussion could be useful both to the formulation of policy and to alliance politics. That is, it might prevent situations in which the policies of different nations conflict and possibly negate each other, and competition among alliance countries might be reduced. It is thus disconcerting to realize that President Carter's energy initiatives—as were those of President Ford's administration under Henry Kissinger's direction as secretary of state—have been taken largely without prior discussion with U.S. allies in Europe or Japan.

Moreover, the United States has an interest in evolving European and Japanese energy policies. Since the United States itself may be susceptible to pressure put upon its allies, it must be concerned with the energy policies of these countries—a rather ticklish proposition given the sensibilities of sovereign states. This latter difficulty may perhaps be partially offset by U.S. discussion of its own energy policy options in the IEA forum.

New Sources

There are additional areas for intraalliance cooperation and lessened vulnerability through the development of "new" sources of oil.

Current longer-term estimates of highly prospective oil-bearing regions are limited very largely to the offshore. Prominent among them is a nearly continuous belt lying along the USSR Arctic, Alaska, the Western Canadian Arctic, Greenland, and Spitsbergen. It is never possible to estimate productiveness without most extensive surveys and actual drilling; the most that can be said is that based upon present knowledge and assumptions, there ought to be important oil (and gas) resources in these latitudes.[1] This environment has been perhaps the most difficult in which to work in the history of petroleum. The advanced technology for doing so is primarily in the possession of the private oil industry. The latter is, therefore, an asset of incomparable value.

The United States could take the lead in sponsoring an international commitment of the free governments directly involved, a commitment whose purpose is to encourage and facilitate an immense undertaking by the international oil industry, perhaps as consortia, or perhaps as companies acting singly, to determine what is available and to develop discoveries with utmost speed.

It would be an unprecedented endeavor of immense costs, but it would be a clear signal of our resolve to remove at least some of the bonds created by a geopolitics of energy that has so sharply focused our attention on, and tied our energy interests to, the Middle East.

Strategic Reserves

The United States should make a particular effort to insure that IEA members rapidly create truly effective, in-place strategic crude and product reserves. Creation of such a reserve is far more effective when many leading importing states take similar actions. We have warned in this report that stock figures supplied by member states of the International Energy Agency continue to be suspect, partly because the very IEA definition of stocks permits member states—including the United States—to describe their situations in more optimistic terms than the facts probably justify. It is essential that this highly defensive and necessary creation of reserves proceed rapidly, convincingly, and on an adequate scale.

Allies in the Middle East

Reduced U.S. dependence on Gulf oil, which has been

identified as one aspect of a comprehensive U.S. energy policy, would also serve the U.S. interest by involving it less in
the anticipated intense competition in the 1980s among
Europe, Japan, and the USSR for access to Middle East oil.

If we have alternatives, or can create them, the United
States could avoid some of the significant complications certain to rise from the fact that Europe and Japan are highly
dependent upon the Gulf and that the United States is less
dependent. Similarly, there are risks to our alliance's cohe-
siveness if the United States continues to play a leading role
in the oil policies of key Gulf producers. For the latter, the
oil markets of Europe and Japan are so much more important.

Conditions for a more enduring peace in the Middle East
may depend more than the United States admits on a consciously abetted renaissance of European (and a strengthen-
ing of Japanese) economic and political interests in the region
and a consequent diminishment of the seemingly overwhelming U.S. presence. It need not imply that the United States
has any less concern for the security of the region, a concern
that, from both the European and Japanese perspective, is
expressed in a strong belief that the U.S. military commitment is in their supply security interest as well.

Such a course does, however, imply a major shift in U.S.
policy vis-à-vis Saudi Arabia. Caught up in the euphoria of a
spate of opportunistic U.S. energy overtures after 1973-1974,
not one of which appears to have been dispassionately argued
then or since, the United States is viewed from Europe and
Japan as having succeeded in substituting itself in the
monopolistic role in Saudi oil previously held by ARAMCO
and its shareholders. This probably exaggerates what the
United States has really accomplished, and it is not an estimate liked by some Saudis; but it carries a baggage of political implications.

Is the U.S. government acting in a kind of "trust" capacity
for Europe's and Japan's supply? Or as the mainstay of a still
very potent set of U.S. private oil interests? Or is it keeping
a position of great influence over Saudi Arabia to help keep

Europe and Japan in line? Or to be in a position to counter
a Soviet move southward? Or does the U.S. government now
have so prominent a presence in Saudi Arabia primarily in
order to buttress the present regime? Or to help (whom?)
maintain "peace and stability" in the region? Against whose
moves to alter the present balance of power and influence in
the Gulf? Iran's? Or, by being in Saudi Arabia, does the U.S.
government have an additional fulcrum against which it
might be able to exert pressures—in behalf of Israel—in the
Middle East? On Europe? On Japan? Or is the U.S. government in Saudi Arabia for any or all of these reasons, reasons
that may be supportable but that have not been carefully
weighed in a process of disciplined inquiry?

This inquiry should have the highest priority. The ramifications of these questions extend deep into our key alliance
relationships. If we explore them, we are bound to come up
against the portentous fact that we are reliant in varying degrees on the Gulf for a commodity that is vital to us severally
and collectively. The starting point of such an inquiry should
be to define the range of U.S. purposes for having so large a
presence in the oil of the Gulf. The conclusions of the inquiry
then have to be supported; it must be ascertained that the
American public supports this role. Finally, it must be ascertained that the defense and other commitments to support
the role will be made.

If the United States does not itself possess sufficient direct
and vital interests, or if there are alternatives, it is risky to
contemplate a role simply for the purpose of influencing
allies on matters in which their interests are substantially
greater than our own.

The outcome of an inquiry might result in the United
States making plain to allies what is now not clear: that while
the decision is the Saudis', it is not the U.S. intention to have
an exclusive relationship with Saudi Arabia; that, in actuality,
the United States reaffirms its belief in the general beneficial
effect of an "Open Door" to the oil of the region, the oil of
Saudi Arabia in particular; and that the United States intends
to help maintain the stability of the Gulf and help assure

security of supply. Exclusivity almost invites the very occurrences one hopes to avoid.

None of these several initiatives will result in a total identity of views or in the establishment of complete harmony within our alliances; nor are these the only objectives (one of which is to decrease U.S. vulnerability to the geopolitics of energy). The different resource positions of Europe and Japan compared with that of the United States will continue to make for divergent interests and differing world views. Moreover, countries will retain the right to go outside the energy alliance of the IEA directly to producing states. There is also the possibility that U.S. allies will move cautiously in going beyond the realms of the alliance; bilateral arrangements could have costly economic and political consequences.

Energy and the "North-South" Dialogues

In the aftermath of the 1973 OAPEC oil embargo, it became increasingly clear that access to raw materials would no longer be determined automatically by the needs of the industrialized world. Access to raw materials could in no way be considered an acknowledged "right," derived from a defunct colonialism or its remnants. By repeatedly linking the issue of access to oil to the demands of the less-developed countries, the oil-producing states have guaranteed that developments in more general "North-South" relations will have important implications for access to energy raw materials. Actual trade, aid, and investment policies emerging from the "North-South" dialogues (presently occurring in many international forums) may affect access to energy raw materials, but, on a less formal level, the changing relations among countries occasioned by the "North-South" dialogues may also have important implications for access to energy (oil and uranium).

In the immediate term, the oil producers have repeatedly warned that lack of progress by the industrialized world in meeting the demands of the LDCs will result in higher oil prices. Most recently Sheikh Yamani, the Saudi oil minister, has warned if the West does not make concessions to the LDCs, oil production may be cut back. Regardless of the sincerity of the OPEC commitment to the LDC cause (and it varies from OPEC member to OPEC member), having made the commitment, it would be politically embarrassing for OPEC to renege on it now.

Moreover, it costs OPEC very little to champion LDC interests. The happy possibility of raising prices to punish an "unrepentent" industrialized world hardly seems onerous. Yet OPEC-LDC solidarity should not be overstated. With economic recovery in the West and Japan, higher oil prices were probably in the cards in any event; consumer intransigence is merely an additional justification for a price increase that may have been coming in any event.

This is not to downgrade the oil link. In fact, it seems quite likely that the unwillingness of the oil producers blatantly to betray the LDC cause does put pressure on them to do something to demonstrate their continued support of the oil link. The rising cost of the link (LDC demands for increasing aid from the oil producers) also motivates the producers to accomplish something for the LDCs, something that will take the producers off the hook. Concessions by the industrialized states to the LDCs probably strengthens the position of the OPEC "moderates," who argue for price restraint and production adequate to meet Western and Japanese needs. Consumer intransigence probably stengthens the position of the OPEC "radicals," who demand higher prices (either directly or through the more provocative mechanism of production cutbacks). Under the circumstances, continued intransigence makes it harder for OPEC to continue to meet world oil supply need as demand rises.

However, the "North-South" dialogue, which actually took off in December 1975 with the convening of the Conference for International Economic Cooperation, was meant to have more long-term implications for world oil trade. The oil link was to establish a three-way relationship in which

concessions to the LDCs would be reciprocated by an undertaking with the oil producers guaranteeing secure access to
adequate and continuous oil supplies at "reasonable" prices
(again, from the perspective of the industrialized countries).
Such an agreement appears unlikely. Having only recently secured the political and economic power to determine prices
and production levels, it is difficult to understand what
reason OPEC would have for sharing that power with the industrialized countries.

Moreover, OPEC's demands on its own account (security
of financial assets from the effects of inflation and currency
depreciation, investment opportunities secured against political risk, greater influence in international monetary affairs,
and assistance in accelerated industrialization) will be agreed
to by the West and Japan out of their own self-interest, regardless of an oil agreement, or so OPEC will argue. From the
producers' perspective, attractive investment opportunities in
the developed countries, hedged against inflation and currency
devaluation, are necessary to (1) encourage those oil-exporting countries that are producing oil in quantities that generate income beyond their immediate revenue needs to continue to do so; and (2) encourage the petrodollar recycling
necessary for balance of payments stability in the industrialized countries. "Indexation" may continue to be rejected
by the developed countries—however much they practice it in
their domestic economies—as being inflationary and also as
increasing the prices of raw materials whose producers do not
now (and may never) have the ability to control prices and
production unilaterally.

In short, there is little the developed countries have to
offer the oil producers beyond those things that the industrialized countries may have to offer anyway for reasons of
their own self-interest.

The relationship between producers and consumers was
meant to be more complex; if the industrialized nations had
little to offer the oil producers directly, they did have assets
to offer LDCs in general. However, Western and Japanese
concessions to the LDCs will not result in an oil agreement

that guarantees oil supplies or stabilizes prices; OPEC has little interest in such an agreement. This is particularly true if the oil demand-supply situation will be increasingly tight; under such circumstances OPEC will not want to be bound by an international agreement.

If an international agreement securing oil supplies is unlikely, what then is the relevance of the "North-South" dialogue for access to energy? In the first place, it is conceivable that some different arrangements for commodity trade might be agreed upon that could have an impact on oil. More importantly, however, international relationships are changing, and access to raw materials, including energy, may well turn on the kinds of relationships that emerge between individual countries and blocs of countries from the "North-South" dialogues.

The U.S. government opposed the convening of CIEC with its links between issues and aid, development, raw materials, and energy. The initiative of Saudi Arabia and France to hold the meeting compelled it to participate. It appears that the United States sought from the outset to break the LDC ranks and to prevent any substantial agreement in any of the various pieces of the "North-South" dialogues; such an attitude may be against the U.S. interest.

As for the Europeans and the Japanese, U.S. intransigence saves them from an agreement that may prove costly to these resource-deficient countries, but it does not prevent them from making their own agreements with the LDCs. Such arrangements are not necessarily detrimental to U.S. interest abroad, but they certainly may be detrimental (the negative impact of the EC/LDC association agreements on the growth of U.S. exports is an example). To the extent that delay, apparent confusion, and U.S. policy make it unnecessary for these countries to pay higher prices for a broad range of goods, the Europeans and the Japanese are still free to attempt a more accommodating approach in other ways, even to the exclusion of the United States.

For example, the Lomé Convention (1975) between the EC and forty-six African, Caribbean, and Pacific countries certainly suggests a greater willingness on the part of the EC countries to meet the LDCs, at least halfway. Generalized, nonreciprocal trade preferences, greater multilateral financial assistance, and the innovative STABEX scheme for compensatory financing to stabilize LDC commodity export earnings are symbolic of a more accommodating attitude on the part of the Western Europeans. One hope is that the goodwill resulting from this European willingness to meet the LDCs halfway will spill over into the oil and energy areas.

If an international agreement is impossible, oil import-dependent states may fall back on bilateral arrangements and special relationships, and in this case the goodwill generated by a readiness to deal with the LDCs could prove decisive. The United States, on the other hand, may be handicapped by the LDC conviction that the United States is considerably less than an honest and committed participant in the "North-South" dialogues.

This is not to suggest that the industrialized states should succumb to OPEC pressure and resign themselves to LDC demands. But these LDC demands are not new, even if attention has focused on them only as a result of the oil link. Hard bargaining in many international forums and the arrangements finally agreed to will contribute to the kind of international environment in which future questions of access to raw materials and energy resources will be determined. Moreover, the current environment is conducive to internationally agreed change. The LDCs seem less ideologically militant and more pragmatic. They have implicitly recognized that the West is more essential to their own economic well-being than the Soviet bloc (a point that could not have escaped Soviet notice, although it is not easy to suggest the form the eventual Soviet response will take). The developed countries continue to bargain from a position of strength. With the international environment more conducive to negotiations, which offers the United States a wider range of options, the opportunity must not be allowed to slip. If so, a return to acrimony and confrontation is very likely. While President Carter has indicated his intention to be more positive and forthcoming, evidence of a change is found not only in speeches and interviews but in commitments—and these have yet to come.

What then of the institutional arrangements of the "North-South" dialogue? The dialogue continues in many international forums. The CIEC was due to wind up in mid-1977. It is important to the United States because it is the sole international forum of which the United States is a member that deals with energy issues and that includes both producing and consuming countries.

Because of its scope and composition, the CIEC Energy Commission is a forum for increasing mutual understanding and appreciation of the respective interests of producing and consuming countries. Such a forum is both important and necessary. While it may prove impossible to convince OPEC or the nonoil LDCs that such a limited initiative should continue regardless of developments in the wider CIEC undertaking, an organization separate from CIEC but similar in composition to the CIEC Energy Commission will be difficult to create because it would look as if OPEC were abandoning the oil link and the LDCs, although such a separation would suit current U.S. policy.

Nevertheless, whether part of CIEC or independent, the United States should make an effort to convince other countries of the continued importance of the Energy Commission or the need for a similar organization. Some moderate U.S. cost might be incurred to keep this important initiative alive. The LDCs are aware of the importance of the Energy Commission and may view its continuation beyond CIEC as an LDC concession requiring a like concession from the developed countries. In this regard, the London Summit's agreement "in principle" on the creation of a common fund to stabilize LDC commodity-export earnings may be sufficient to permit OPEC to continue the Energy Commission with at least the apparent winning of the common fund "concession" from the developed countries.

A central purpose in the dialogues is to find ways of meshing the economic interests of producers into the world trading and financial system. Such purposeful efforts may be the only way of diminishing the effects of the geopolitics of energy for as long as we are dependent on oil.

Balanced Interdependence

Some greater protection against the vulnerability derived from the geopolitics of energy can be found in deliberate efforts to redress the imbalance in interdependence occasioned by the unilateral political power of producers over oil production levels and prices. The United States is not without assets and options in this regard.

With specific regard to oil, producers could be encouraged to participate in all functions of the industry. By multiplying their stakes, producer countries may develop interests closer to those of the consumers, even if their greater involvement in oil operations also enhances their ability to manipulate supply and price for other than commercial reasons. Downstream investments in consuming countries might give producers an additional incentive to supply such projects.

There are other means for tying producers' interests to consumers' interests, and policies could be devised with this end in mind. Others have stressed that increased direct and other investments by the producer countries in the developed countries might provide one route. (The key sector approach, which is employed by almost every country to insure national control of sensitive industries, can be implemented as a quite reasonable and widely accepted safeguard.) In this way the general economic health and well-being of the industrialized countries becomes of more than passing interest to the producing governments. Some of them have already quite explicitly recognized that their own economic well-being is intimately associated with developments in the Western economies. Efforts to make this increasingly clear and valid should be intensified.

The United States should continue to reinforce trade ties with an expansion in its exports of goods essential to the producing countries—including food. As development plans are implemented and expectations rise, it will be more difficult for the producing countries to do without imports from the industrialized countries; Colonel Qaddafi's threat simply to return to the desert will be increasingly difficult and unpalatable.

These are the traditional means, the means whereby the interests of societies become mutual and, therefore, durable. Arms sales cannot achieve these results. At the most, they may for a brief period bind some of the interests of one nation to some of those of the other. They can more easily complicate relations among states than create common denominators of national interest. An effort aimed at "buying" oil by arms sales is almost the antithesis of what is recommended.

In addition, there are still exporting country demands that may not be unreasonable and that still have not been addressed by the industrialized countries. Indexation of price is one of them, although Saudi Arabia probably does not want to see the introduction of any automatic mechanism for price determination, since this would detract from Saudi ability to influence OPEC. Moreover, from the U.S. viewpoint, the onus for price increases should remain with OPEC and not be transferred to inflation in the industrialized countries.

Demands for investment guarantees seem reasonable, and if some could be generalized to protect all foreign investors, U.S. foreign investors might benefit as well. Desires for investment instruments hedged against inflation and currency devaluations do not seem out of line, but the banking community would be better able to judge the effects of a more widespread use of such instruments. To date, there is little to indicate movement in these directions, or even serious study of what is a frequently expressed producers' objective. Joint investment opportunities should also be explored. The question of whether countries with government energy companies have an advantage over the U.S. industry should also be appraised, with care taken to differentiate "efficiency of supply" from the issue of presumed "political advantage."

The industrialized countries do have assets and opportunities to redress their energy imbalance without necessarily increasing the vulnerability derived from dependence on imported energy resources. Even so, nothing precludes the producers from taking what appears to the outsider as "irrational" actions; but the costs of such action would be higher.

"Special Relationships" and Preferential Sources

If secure, adequate, continuous, and reasonably priced oil supplies cannot be obtained by international agreement, consuming countries may fall back on bilateral arrangements and "special relationships." The questions to be addressed include: (1) to what extent does a special relationship enhance security of supply? (2) at what cost? (3) under what circumstances do such relationships endure? (4) what opportunities for forming special relationships does the United States possess? (5) what are the problems, perhaps special to the United States, that our governmental–private sector system raises with a "special relationship"? (6) what are the consequences of a special relationship for the particular producing governments, and are those consequences consistent with the U.S. objective of attaining greater security of supply? and, (7) what is the effect of such relationships on third parties?

Historically, special relationships have taken various forms and served various purposes. Colonialism was a form of special relationship that secured for the industrial state any number or combination of objectives: (1) areas for settlement; (2) access to raw materials; (3) potential export markets; (4) investment opportunities; (5) sources of labor; and, (6) the trappings of power in a world in which competition for colonies was intense and colonial possessions symbolized strength. By and large, colonialism represented an almost exclusive relationship between the metropolitan country and the colonies.

With the increasing drive by colonies for political independence and the weakening of empires in two world wars, the interests of each side became sufficiently different to make unlikely a continuation of the relationship on the same terms. With the rise of the United States and USSR on the world scene, the colonies had alternatives to the "mother" country, and the ascending powers were eager to break the exclusivity of the colonial relationship. Colonialism was an artificial creation dependent on the overwhelming power of one side to maintain it and the weakness of the other side to resist. Colonialism could not survive the changes in its foundation.

Or, special relationships may emerge for naturally occurring causes, for reasons of a real coincidence of national interests, or for reasons of deliberate government policy (or some combination of these).

Geographic location naturally influences relations among states. The Soviet Union has special relationships with the countries of Eastern Europe. While there are many reasons for this, at least one important reason must have been the Soviet desire to protect its western flank from a resurgent Germany. Moreover, geography has given all these states a mutual interest in seeing that Germany will not threaten them again by providing for a collective defense that could deter or defeat a resurgent Germany. (The latter point, of course, was an original cause binding NATO allies together in their special relationship to West Germany.)

Such special relationships, based on the foundation of common interests, prove most durable. The Atlantic alliance is an example of a special relationship based on shared common interests. It endures because, regardless of differences in relative power, world views, and judgments as to appropriate means, both the United States and Western Europe believe that their fates are inextricably tied to each other.

With regard to access to oil, the central question is the capacity of the United States and hemispheric oil producers to create by government agreement a special relationship that will reflect the mutual interest of the participants. Under normal circumstances our system would look naturally to the private sector to accomplish this purpose. The history of "foreign oil" in all the areas under discussion virtually precludes our customary approach: political reasons embedded in their national experiences will not allow it.

A balanced consideration of the value of special relationships will also weigh the costs of such relationships and the issue of increasing dependence on single sources, even if those sources are not Arab. An important element in cost considerations is the degree to which the special relationship commits the United States in terms of price. If the oil demand-supply situation should change, the United States will not want to be bound by prices that no longer reflect prevailing market conditions.

It is necessary to define the questions that must be addressed if a policy of cultivating "special relationships" and preferential sources is to be intelligently evaluated. The opportunity for developing such relationships may actually be at hand. Special relationships with close neighbors offer the United States one major avenue for reducing its dependence on Gulf oil, a goal quite as important for its allies as for itself. But there will be costs.

Further study could profitably deal with all these issues in far greater detail, including the question of costs and increasing U.S. dependence on hemispheric sources of supply. It may well be that oil and energy are so very important that cost is a less relevant consideration than security of supply, but the cost should be known. One must also not assume that an import source is likely to be "secure" until one has weighed several considerations, such as geographic diversity, the adequacy and potential of resources and producing capacity, the existing range of interests of the producer in its relationship with the United States, the comparative security of its sea supply lanes, the extent to which the exporter may have an interest in issues—such as Israel—that could cause it to participate in an embargo, and its need for revenue.

Venezuela

During the 1960s Venezuela repeatedly indicated in a

number of ways its interest in preferential access to the U.S. oil market. That is, in return for preferential access to the restricted U.S. oil import market, Venezuela would guarantee quantities of oil supply. At the time, U.S. government officials virtually ignored the Venezuelan proposals. Private oil companies, involved in negotiations with the Venezuelan government regarding their concessions, cautioned the U.S. government against any overtures to Venezuela that might jeopardize the negotiations and their interests. In the event Venezuela became a founding member and staunch supporter of OPEC, the concessions were eventually lost, and the United States may have missed an opportunity to guarantee supply. The old concessions are securely under Venezuelan control, and there is little the United States can offer in this area; the Venezuelans, in new relationships with the private oil companies, need no additional assistance in this area of oil production.

The same cannot be said of the Orinoco petroleum belt. Here U.S. financial and technological assistance could enable the Venezuelans to tap this potentially huge source of oil (and oil revenue). Having only recently nationalized the concessions of the major international oil companies, it may prove politically difficult for the Venezuelan government to invite these same companies back into an important position in the Venezuelan oil industry by giving them development responsibilities in the Orinoco.

A U.S. government initiative designed to assist Venezuelans in developing the petroleum belt may be less objectionable in terms of domestic Venezuelan politics. However, because the United States is hardly more palatable than the major international oil companies, which were viewed as agents of the U.S. government in any event, the terms of such U.S.-Venezuelan cooperation must be firmly rooted in Venezuelan interests and must scrupulously avoid any semblance of exploitation.

In spite of advances in transportation, the United States remains a natural market for Venezuelan oil and oil products. It could then assist in the development of the Orinoco with reasonable assurance that any product from this area would naturally find its way into the U.S. market. The United States, therefore, could refrain from demanding such a guarantee but could encourage the eventuality by offering preferential treatment to oil from the Venezuelan petroleum belt. As the United States is Venezuela's natural market, however, a preferential agreement may not be necessary.

Consideration should also go to two additional questions: (1) how does the Venezuelan commitment to OPEC solidarity affect the possibility for such a U.S.–Venezuelan relationship? (2) since the United States is a natural market for Venezuelan oil and since U.S. withdrawal from the Middle East oil market is as beneficial to the Europeans and Japanese as it is to the United States, might not U.S. allies be encouraged to assist in the development of Orinoco and thus make the initiative more acceptable politically?

Mexico

The Mexican government nationalized the operations of the private international oil companies in 1938. In spite of the fact that this was almost forty years ago, it is still viewed with great national pride as marking the real beginning of Mexican independence. In addition, the Mexican government established a state oil company, PEMEX, which by now has accumulated considerable experience and skill.

The possibility that offshore Mexico may hold large oil reserves and that the Reforma field and others might represent a major oil field discovery, and the consequent possibility that Mexico will become a more significant oil exporter, make a definition of U.S.–Mexican relations important.

PEMEX may be capable of developing Reforma alone, in which case an agreement on preferential access to the U.S. market might be reached, but here, too, the issue of preferential treatment must be weighed in light of the fact that the United States may be a natural market for Mexican oil. Of course, nothing really prohibits the Mexicans (or the Venezuelans for that matter) from incurring the transportation cost of shipping Mexican oil to Europe, and the issue of preferential treatment must be carefully evaluated.

Should the Mexicans reject assistance in oil development, there are a whole host of additional Mexican interests that could be addressed in a special relationship—trade, investment, labor issues, and the like. A special relationship of this sort must necessarily cover areas other than energy raw materials, and the cost to the United States may therefore be higher than a straight oil arrangement. Making such a relationship politically acceptable to Mexico may prove even more diffi- cult than in Venezuela, which is not to suggest that an effort in this direction, requiring more skill than the U.S. govern- ment has previously shown in its dealing with LDCs, should not be forthcoming.

As far as the United States is concerned, the choice is not either-or; i.e., selecting Venezuela for a special relationship does not preclude a similar relationship with Mexico. In fact, if a special relationship materializes vis-à-vis Venezuela (or Mexico), the other country may be more willing to negotiate a similar arrangement. If both countries participate, the special relationships might be more politically acceptable.

Canada

The United States has long had a de facto special relation- ship with Canada in the energy area. At the same time that Venezuela was requesting a special relationship with the United States, the Canadians received special consideration in U.S. energy policies. The relationship evolved from the close intermeshing of the economies of the two countries, and the special situation of Canada often required special Canadian exemption from U.S. laws. It is an understatement to note that while recognizing the benefits of the U.S. tie, the Canadians were hardly pleased with their dependence on the United States. Particularly in the area of natural resources— including oil—the control of these resources by foreigners (largely American) is an irritant in Canadian-American relations.

In the aftermath of the 1973 embargo, the Canadians took to opportunity to claim Canadian oil for the Canadians. The oil export tax was raised, and exports to the United States were affected. These exports are to be progressively reduced as the Canadian logistics system is extended to allow greater interprovincial flows of oil.

The United States and Canada might develop joint projects of interest to both countries—including exploration and development in the Canadian Arctic, U.S. assistance in developing Canada's extensive tar sands deposits, and perhaps more cooperation in the nuclear energy area. We may seek with Canada joint efforts in gas transmission; we have agreed upon the terms of a general pipeline treaty. Moreover, the Canadian government now accepts the certainty that it will become an energy importer as its current sources of oil become depleted. As the full import of this assessment sinks in, we expect fresh efforts by Canadian private and public companies (PETROCAN) to explore on a wide scale. The United States—privately and publicly—may find fresh opportunities to assist.

Saudi Arabia

We have returned repeatedly to the pivotal role of Saudi Arabia and the continuing need to adjust U.S.–Saudi relations in light of Saudi Arabia's increasing importance as the major oil producer. With spare capacity in a tight oil demand-supply situation, Saudi influence in OPEC, and the essential importance of Saudi oil supplies to U.S. allies, questions of U.S. relations with the Kingdom are central.

Saudi decisions on production levels and price will be of exceptional importance. A recent study has received widespread notice, for it emphasizes the necessity of an early Saudi decision among three choices as world demand rises to meet available supply. The formal limit now placed by the

Saudis on their production (8.5 mmb/d) is some 3 mmb/d below present capacity. The margin is the crucial factor. According to the study, the practical, available Saudi capacity surplus may be 1.5 mmb/d, but, most importantly, it may represent as an actual matter some 50 percent of the true surplus capacity in OPEC's whole membership.

The Saudis have three choices as they decide whether to meet increasing world demand: (1) maintain the production limit against the increasing demand, (2) lift the limit, and (3) lift the limit and raise prices. Any one of these would be a signal demonstration of Saudi oil power. (In the last OPEC conference they chose the third.)

There is a fourth choice, which may be equally attractive. It is to *lower* production on the grounds that Saudi oil revenue is wildly excessive to its needs, that there has been no progress on its insistence that excess revenue invested abroad be protected against loss in the value of the dollar (or other currencies), and that there is no durable, reasonable rate of return for Saudi Arabia's diminishing resource.

This decision, which may be as likely as any of the others, would have several immediate effects on the revenue earned by all OPEC states, since a price increase would be inevitable. (It would also, of course, bring an increase to Saudi Arabia, but the effect on all other OPEC states would be great and could, indeed, be one of the reasons in support of such a Saudi move.) The other effect would be ominous: a lowered production level would bring producers and consumers closer to the crunch—where demand soon bumps against supply, and where the pressures to get oil and thus the danger of miscalculations would build.

When all these possibilities exist, it is extraordinary that neither the United States nor anyone else is thought to be knowledgeable enough about the Saudi elite to have a more convincing appreciation of the probable directions of its oil policy. For all of the U.S. vaunted "special relationship," others are uneasy over the signals one Saudi or another flies; they suggest that the United States has either misled itself into thinking that it knows the Saudi direction, or the Saudis may have misled the United States. Perhaps it is both. Perhaps there is no unity among the Saudi elites on these positions, or perhaps the intricacies of the Saudi role in world oil are only now being understood by influential cliques within the royal family. The blunt fact is that we do not seem to know, which makes judgments about the directions and durability of the regime more like guesswork.

This is the place, then, to be reminded of the principal observations usually made of the Saudi government:

1. The members of the royal family are close to one another, and there is little prospect of a serious division of interest, which would be an opportunity for a coup;

2. Even if there were a coup, the Saudi interest in oil revenue would compel them to maintain the flow (and it would be a good idea immediately for a new regime to assure the United States, Europe, and Japan that this is to be the case);

3. The Saudis cannot generate among themselves a skilled (and unskilled) labor force large enough to put even a scaled-down version of their development program into effect; imported labor on a very substantial scale will continue to be needed;

4. Similarly, with management skill;

5. The Saudis will probably not be able to meet their principal development goals in time, and the frustrations attendant to this anticipated failure will be a serious factor in internal clique alignments;

6. As long as the golden eggs are distributed generously and with a sensitive hand, the chance of disaffection or envy "getting out of hand" is remote;

7. There is little prospect that the Saudis could defend themselves against an attack from outside (Iran or, conceivably, Iraq); Saudi dependence upon U.S. support of all kinds is thought to be an essential ingredient in Saudi thinking; and

8. Saudis prefer to deal with Americans for a variety of reasons, including their experience with ARAMCO.

What is curious about this list is that it does not contain serious warnings; read in toto, it seems to assure stability.

But applied to another country, the list would be regarded almost as a prediction of trouble: a vast country of untold energy wealth that is vital to the needs of great, industrial powers; a regime based upon an extended, almost tribal system held together to some degree by loyalty but even more by its access to Croesus-like wealth and its "fair" distribution; a country unable to defend itself without dependence on a major external power; a country moving into modern times at breakneck speed with all the superficialities one would expect, dependent to an inordinate extent on large numbers of imported laborers and managers; a country located in a region where historic issues and relationships seem to tend to divisiveness and suspicion.

These observations are much needed reminders of the weak underpinnings of a group to which the United States is so deeply committed and for whose conduct in oil matters the United States will be held partly responsible. This is the situation—an extremely difficult one—in which the United States and Saudi Arabia find themselves, or have put themselves. Whether the relationship is durable or subject to swift change is, of course, a key question. Whether the United States, anticipating difficulties in Saudi Arabia, in the Gulf, and with European allies and Japan over its almost exclusive position in this oil giant, should modify its relationship has already been discussed.

The Soviet Union and the People's Republic of China

As for the Soviet Union, unless it has easy access to the capital and technology and management skills of the oil industry, we do not anticipate that it will be able to be a major oil exporter to the industrial world within the foreseeable future. Instead, it is more likely that the USSR will have to import oil—probably from the Middle East—to meet its domestic needs until it is able to draw upon the prolific fields said to be in Siberia. As such, the USSR may not be a serious factor directly complicating Western and Japanese oil supply.[2] As a competitor for relatively small amounts of Middle East production, the USSR probably could not use oil as a disturbing influence in the region. Under the circumstances, U.S. policy will have to assess carefully the desirability of providing technical and financial assistance to Soviet energy development.

We have warned repeatedly, however, that the situation prevailing for oil may not hold for natural gas, where the potentially leading suppliers in world trade are likely to be the USSR and the Middle East. It is not so much a case of their gas becoming a major source of energy in Europe and Japan as it is a danger that its selective end-use in the industrial economies could give crucial leverage to the suppliers.

In the case of Communist China, its energy potential is impossible to discuss intelligently in any detail. No one knows, although some claim to do so. Based on the fragmentary evidence accumulated, it is more likely than not that China will have its hands full with the priorities and requirements of an economy moving into the petroleum age. Such oil as may be made available for export would not be enough to diminish significantly Japanese dependence upon the Middle East, for example. In short, we have not considered China itself an important factor in energy world trade.

14
Nuclear Energy

On the nuclear fuels supply issue, the United States must move on several points. But it must move not from a sense of adequate endowment of uranium, for there is no reasonably persuasive evidence that this is the case. On the contrary, given the possibility of a shortage of ore and enrichment facilities, and given the probability that the present lead of U.S. technology will diminish, the United States can exercise nuclear "muscle" no longer than the next decade or two at the most. Other countries will not wait for the ore shortage to come about or for the technology gap to close; they will undertake their own initiatives to assure supply. As these developments are anticipated and as the long lead time needed is recognized, new directions will be charted in nuclear energy acquisition. We have seen evidence of this in many places: Germany, France, Spain, England, Japan—and links to Brazil, Iran, and Pakistan.

Nevertheless, the intimate relationship between nuclear energy for war and peaceful purposes compels the United States to move with great firmness toward the adoption of its concept of international, regionally located fuel enrichment and other processing facilities. The difficulties lie in three directions: (1) the interest of Germany, France, the U.K., Japan, and U.S. government and industries to develop major stakes in the provision of commercial nuclear plants; (2) the interest of nuclear-to-be industrial and developing states, which see in their capture of the whole fuel cycle both the symbol of modernity and the means for weaponry; and (3) the potential uranium ore–exporting nations—such as South Africa, Australia, Canada, and Gabon—who have every reason not to export the ore itself, but to export it as fuel. For all these varied interests, the otherwise obvious attractions of regional processing centers begin to fail.

The convening of the "Suppliers Club" of the leading nuclear countries has been presumably to define the terms on which nuclear energy for peaceful purposes is to be made available to LDCs (and other industrial states). There are those, of course, who ask whether this is not another OPEC-in-the-making, whether, in view of the apparent attempt of uranium ore producers to conspire, this is not a comparable undertaking. These assertions raise many complications. The one we single out is both a warning and an alert that an opportunity may be seized: it seems quite likely that nuclear energy will become a topic in energy discussions previously limited to oil; the demand for one will be linked to the other. If they carefully and imaginatively handle the matter, some nations could forge a constructive link that would put energy generally, and not specifically one source or another, into a more general world economic setting.

Surely it should be possible for a number of regional fuel centers to be equipped by suppliers who compete only on the reactor level. If it is not possible to attract ore-exporting nations to the regional centers, then it may be necessary to examine the potentialities for "special relationships" between the United States and exporters in order to assure the United States of adequate uranium ore. We know that other states are actively considering such commitments. Most recently, the Japanese government has undertaken to explore and develop Indonesian uranium supply potentials. Japan has similar intentions with Canadian uranium resources. The examples will multiply. If so, most of the observations pertinent to special relationships for oil are valid for uranium.

Finally, the creation of national banks of uranium ore, most advantageously after some degree of processing, is as fundamental a security need as is an oil strategic reserve. The difficulty is also analogous. If each nuclear power attempts to create a strategic reserve all at the same time, the available supply may not permit it.

15
Epilogue: Policy Recommendations

In order to reduce its vulnerability to possible emergency situations, the United States should:

1. Maintain the International Energy Program as a symbol of the unity and commitment of the industrialized countries. A continuing effort is necessary to strengthen the International Energy Agency by defining its functions and commitments in more realistic ways than has been the case to date.
2. Establish with utmost urgency the Strategic Reserve, which has already received Congressional authorization and now requires intensified effort. In order to enhance the value of such a reserve, U.S. allies should be encouraged to establish their own strategic reserves. Additional work is still needed on defining and measuring adequate strategic reserves.
3. Establish a watch over tankers, particularly the future availability of the smaller tankers, which continue to play the major role by far in U.S. import supply. Product tankers generally need to be watched as well, since more and more oil in world trade will take the form of products rather than crude.
4. Establish a watch over refining capacity, since U.S. excess capacity (when Caribbean refineries are included) may well vanish and since U.S. products demand, without new refinery construction, may have to be met from

European refineries. In addition, if the OPEC countries
do enter the refining phase in a big way, the European
refineries themselves may be "uneconomic": the producers will insist that consumers take their products
first.

In addition to these "defensive" measures, more fundamental, long-term actions should be taken to reduce U.S.
vulnerability in this respect. Actions designed to limit the
need for imported energy supplies and to secure access to
those supplies must be taken.

1. Of primary importance is still the need for a comprehensive National Energy Policy and Plan, which implies the
need for a greater government role in coordinating, oversee-
ing, and encouraging energy development in directions consistent with the U.S. national interest while reducing the prevailing uncertainty, which discourages private initiative. Giv-
en the interrelationship between all aspects of the energy
equation and given the need for actions to occur in a timely
manner, a comprehensive National Energy Policy is the first
priority. President Carter has urged this, and most of his proposals echo those of the previous administration.

2. But in order to implement a National Energy Policy,
centralization of governmental energy functions and responsibilities is essential. Equally important is the attention that
must be given to reform of the regulatory procedures that
have caused confusion and needless delay. Conflicts of energy
"jurisdiction" between parts of the Executive Branch, within
the Congress, and between the states and the Federal Government, must be taken up and resolved. Without success in
these respects, no otherwise intelligent energy policy and
plan can succeed.

3. As for the specific elements of a National Energy Policy
and Plan, most energy analysts as well as Presidents Ford and
Carter have agreed that the following are essential:

a. The intensive development of coal;

b. The search for indigenous and hemispheric supplies of

oil, gas, and uranium ore;

c. The maximization of efforts to bring nuclear energy into the economy on widely accepted terms;

d. The intensification of energy research;

e. The conservation of all fuels and their selective use.

4. The possibility of improved relationships between the U.S. government and U.S. international oil companies must be thoroughly explored.[1] In some areas, where the necessary energy developments are beyond the capabilities of the private sector, the role of government may be essential. The generally adversary government-business relationship in the United States must be compared to the relationship existing in other countries in order to assess the U.S. competitive position. There are other questions as well. Merely to suggest a greater government role says little about the role, capabilities, and opportunities of the private sector, which must remain the dominant actor in energy. Enough has been included already in this review to make clear the essential nature of the private sector even under changing (largely political) circumstances.

If the U.S. government and oil industry are mutually to reinforce each other's efforts, with the government defining the national interest and objectives, the industry must finally adjust itself to these realities. One such adjustment is the government's unquestionable right to know, in advance, of the terms of oil supply agreements reached between U.S. companies and producing governments, and to review, and disallow if necessary, the terms against the stated national energy objective.

U.S. general security interests include the continuing and inescapable dependence of U.S. allies on energy imports. The U.S. must be aware that the other industrialized countries are more dependent and have fewer options, which makes them extremely sensitive to U.S. action in areas where their interests are so much more heavily at stake. These differences make for divisiveness in the alliances and call for greater U.S. responsiveness to the needs of its allies.

1. We believe that the United States has energy options and should:
   a. Reduce its dependence on Gulf oil;
   b. Explore and develop possible oil and energy sources closer to home, particularly with regard to the potential for a particular kind of "special relationship" with Canada, Mexico, and/or Venezuela;
   c. Take the lead in proposing multilateral public-private development of potential Arctic sources of oil.
2. The United States should reassess its relationship with Saudi Arabia, and, while making clear beyond any doubt that U.S. interests are heavily involved in the country, should make equally clear to its allies its intention of maintaining an "Open Door" policy toward a country of such consequence. The Saudis, of course, retain the ability to establish or break relations with any and all nations as they judge to be in Saudi Arabia's national interest. But the U.S. perspective on the relationship, in the light of the interests involved, needs to be clarified.

The issue of access to oil is closely related to developments in "North-South" relations in general and to the new international relationships evolving from "North-South" dialogues.

1. It is therefore necessary to reassess the U.S. position in this arena. The continued conviction of LDCs that the United States is the major stumbling block to a new international economic order is probably not in the U.S. interest in a world where access to raw materials and energy raw materials will be determined by evolving relationships. U.S. allies may not be opposed to the United States' self-styled spoiler role (although this may be moderating), since it spares them the need to make costly concessions to the LDCs. At the same time, this role does not preclude Europe and Japan from reaching agreements with the LDCs, agreements from which the United States may well be excluded.

2. Resignation to LDC demands is not our recommendation, but continued hard bargaining based on changed circumstances—in which access to raw materials will not be determined automatically by the needs of the industrialized world.
3. The United States should give generous and immediate attention to the energy needs of LDCs, assisting in whatever ways it can to explore and develop their indigenous energy resources and, through access to research and technology, advance their energy calendar through early opportunities to take advantage of renewable energy resources—solar, geothermal, nuclear, as the case may be.

Finally, with regard to the Soviet Union, which probably enjoys the most favorable energy resource endowment of any of the major powers, including the United States:

1. The U.S. government and private interest should refrain from extending technical or financial assistance to the USSR for the development of Soviet energy resources; such assistance, if granted, has at least one consequence: it would permit the USSR to devote greater attention to other objectives, such as its military programs.
2. The United States should closely monitor the trends in Soviet gas exports to Western Europe and eventually to Japan, as well as the export of Middle East gas to Europe through the Soviet pipeline network. If important segments of the industrialized countries' economies depend on Soviet or Middle East gas and are thus subject to potential Soviet control, the security implications are obvious.

Energy has come of age. It is no longer possible to consider its supply as anything other than a vital national interest. Nor is it still possible to think of it as a commercial commodity. Nor will we be permitted to think of it in largely economic or financial terms. It is all of these, of course, but now it is much more. If the geopolitical approach explains the broad range of concerns that confront societies in their search for an adequate energy supply, then we can fashion an effective National Energy Policy and Plan.

Appendix

U.S. Energy Policy in a World Context

by Walter J. Levy

1. In about three weeks' time, the President will announce the outline for a new United States energy policy, dealing with all aspects of our current and future energy supply and demand position.

I. The Energy Setting for the Next 10 to 15 Years

2. In establishing its policy our government must undoubtedly take into account that there can be no "Energy Fortress America." Any U.S. energy policy can only be viable within our given political, economic, financial and strategic framework if it is coordinated with the vital interests and actions of the other member nations of the non-Communist world. I will, in short order, develop the arguments in support of this intimate interdependence. But let me perhaps first set the scene by giving you as succinctly as possible the world energy picture as it would now appear to evolve during the next 10 to 15 years. The statistics are drawn from the recent excellent report on the "World Energy Outlook" by the Organization for Economic Cooperation and Development.

A. *The Prospects for Oil*

3. We will concentrate our comments on the oil sector, as

oil will remain the major basis of energy supply for the non-Communist world—certainly during the medium-term—and as the major international policy issues will arise in connection with the availability and payments for oil.

Accordingly, I am addressing myself to two questions:

First, is there going to be enough oil available to cover the essential needs through the 1980s and beyond the early Nineties; and

Second, what will be the cost of the oil and what are the balance of payment implications for many of the developed and most non-oil-producing developing countries?

4. If current policies governing supply expansion and conservation are continued, the oil import demand of OECD countries by 1985 would amount to 35 million barrels daily (as compared with 25 million b/d in 1974) and for oil-importing developing countries to some 4 million b/d, or about the same as in 1974. The combined total for the non-Communist world would thus reach 39 million b/d. By 1990, this total would have increased to 45 million. About 90% of the imports will originate in OPEC countries.

5. On the basis of optimistic assumptions for maximum conservation and a massive effort of developing all sources of energy, total oil imports in 1985 would still be 29 million b/d and 25 million b/d in 1990. As the underlying assumptions for arriving at these estimates have been set forth in the OECD Report, I will not repeat them here.

6. To achieve the supply expansion and the demand reduction implied in the optimistic OECD case would pose a most difficult task. Without burdening this presentation with further statistical details, let me just mention that it would presuppose an almost nine-fold increase in atomic energy production between 1974 and 1985—which in terms of oil would be the equivalent of about 9.5 million b/d—an over 40% rise in world coal output—sufficient to replace by 1985 the equivalent of over 6 million b/d of oil—and an advance in U.S. oil production that would presuppose an annual discovery rate of over 4 billion barrels a year—or some 50% above any previously sustained level of discovery.

7. Accordingly, we will base our analysis on an intermediate case reflecting more or less the arithmetic average between the two cases; implying an oil import level of about 34 million b/d in 1985 and 35 million b/d by 1990. We should stress that even those figures still reflect a very optimistic assessment indeed of our future oil import dependence.

B. The Future Cost of Oil Imports

8. The cost of OECD oil imports in 1985—with an inflation factor of 7.5% per annum—would reach in then current dollars some $233 billion and for oil-importing developing countries another $33 billion, making for a combined total of nearly $270 billion. For 1990, the corresponding oil import bill would amount to about $400 billion.

II. Major Issues Implied in the Oil Supply Picture

9. Let me now review some of the policy implications that are posed by the forecast just given. There is first and foremost a real question whether enough oil can be produced for the period beginning around the early Nineties.

A. Physical Limitations on Future Oil Production

10. Let us assume a future average discovery rate of some 15 billion barrels of new oil a year, a figure that is somewhat higher than previous experience suggests. Given this optimistic assumption, the ratio of the world's current oil production to proven reserves would decline from some 33 years as of now to about 20 years by 1990—with the ratio for OPEC countries declining from 40 years to 27 years. Less than one-half of 1990 production would be obtained from new discoveries, the balance would come from reserves known today.

11. Three major conclusions emerge. First: Towards the beginning of the Nineties, even ignoring any restrictions oil-producing countries may impose on their level of output for political or economic reasons, we may in fact be approaching a time when physical limitations will impinge on maintaining, and even more so, on increasing world oil production. This applies to most oil-producing nations, except perhaps Saudi Arabia and a few Sheikhdoms where almost half of the world's known oil reserves are located. Because of the long lead time from exploration to development, the supply availabilities through much of the Eighties are already largely determined; the only major opportunity for obtaining massive new supplies depends on continued free access to known Middle East reserves.

Second: We may soon be confronted by a severe competition among the oil-importing countries for access to oil supplies that may translate itself into policy and strategic as well as economic conflicts.

And third: When oil reserves begin to trend downward, toward a level where current production could no longer be sustained, oil prices would not only respond to inflationary factors, but the real price of oil would also rise and may sooner or later skyrocket.

12. As the national energy demand, the local energy supply and thus the oil import requirements of all the members of the non-Communist world are interdependent, the U.S., as the single largest energy user and oil importer, plays a particularly significant role.

13. What the U.S. can achieve in the field of conservation is of worldwide relevance, not only because it would affect U.S. oil imports, but also because it could provide the technology and would set a pattern for progress in conservation elsewhere. The same observations hold true for U.S. research and development efforts in the field of energy supplies from all sources.

14. A coordinated energy research and development program including all the resources of the non-Communist world is absolutely essential. A future energy crisis can—if at all—

only be avoided or mitigated if such a maximum program is carried out, and if we are lucky and skillful enough to discover in time very large new hydrocarbon resources, or are able to accomplish dramatic technological breakthroughs for new sources of energy.

15. In this connection, it should be stressed that a substantial expansion of atomic energy would appear to be absolutely essential for most of the OECD countries which, unlike the U.S., do not possess large oil or coal reserves. A U.S. policy of going slow because of the still unresolved dangers of atomic proliferation and waste disposal would, if followed by other OECD countries, lead to an energy crunch which might otherwise have been avoided or at least postponed. And needless to say, that if the non-U.S. OECD countries should proceed with their atomic development programs, as I believe they will and probably must, we would still be exposed to the dangers of proliferation and waste disposal, because the atomic threat knows no boundaries. Any U.S. policy of "atomic abstention" would not only tend to reduce or delay the chances of coping with atomic dangers, but would also affect our international influence and standing in this field.

B. Political and Economic Limitations on Future Oil Production

16. Long before limitations on the physical availability of oil would occur, producing countries might impose politically or economically motivated restrictions on the level of their oil production and exports. Here, too, the U.S., as the leading power of the non-Communist world, is best placed to assure the unimpeded flow of oil in the world's trade.

17. It is the U.S. that has the capability to provide protection for the major oil-producing areas against external as well as internal threats; and the relevant producing countries know this full well, whether they strive openly for a so-called special relationship with the U.S. or take it tacitly for granted. Likewise, only the U.S. has the power potential to

secure the availability of OPEC oil for world markets in case internal, regional, or other external forces threaten to interdict its free flow.

18. Also, oil exploration, production and exports rely in most of the major producing countries on the technical, managerial and marketing competence of American oil companies; and overall economic development is substantially dependent on a major input from the U.S. industrial and technical complex.

19. The U.S. thus possesses an indispensable—but not necessarily always effective—position of special influence and bargaining leverage. If the issue of restricting oil exports for political or economic reasons should ever become a threat to the welfare of oil-importing countries, the extraordinary position of Saudi Arabia in terms of the richness of its oil reserves and the abundance of its financial strength is—for better or worse—of particular relevance. The special relationship that has developed between the U.S. and Saudi Arabia provides an invaluable opportunity for exercising a moderating influence. This is equally relevant in convincing the Saudis that in the interest of a stable world, producing countries must adopt a responsible policy in their decisions on oil pricing.

20. This presupposes a willingness of the importing countries to provide sufficient inducements to producing countries to maintain production levels beyond their own economic and financial needs. This could, among others, also involve complex arrangements for the protection of OPEC's surplus financial reserves against erosion through inflation and also other kinds of understandings that would join the interests of the producing countries to those of the importing nations.

21. The creation of the International Energy Agency (IEA)—through the initiative of the U.S.—has eased considerably any threat of a future oil embargo by providing for the accumulation of strategic reserves and for automatic burden-sharing among the importing countries; in addition, the IEA also establishes the institutional framework for a broadly-based cooperation on all major energy problems among im-

porting countries and for the reasonable resolution of issues that may arise between importing and producing countries.

22. Nevertheless, there remains still more than a nagging doubt whether the non-Communist world will be able to cover its future energy needs if all our intermediate efforts do not lead in time to major increases of energy supplies from known or yet unknown sources of energy.

III. The Financial Dimensions of World Oil Imports

23. Let us now turn to the complex problems arising from the balance of payment burden that importing countries must incur for their oil purchases. As mentioned previously, the cost of oil imports of the non-Communist world would reach in current dollars some $270 billion in 1985 and $400 billion by 1990.

24. For the foreseeable future, the non-oil-producing less-developed countries and also a number of the financially weaker developed countries face foreboding problems in covering their current account deficits. Only those few among the oil-importing countries that are financially strong or industrially advanced would be able to benefit directly from the deposits of financial surpluses by OPEC countries, their investments, or their imports of goods and services. Most of the others must rely for an improvement of their foreign exchange position on sustained economic progress or on financial credits and grants they might receive from financially strong countries or international institutions.

25. Much has been said about the past success of recycling, implying that it is a problem that would fade away in due course. In fact, each passing year during which we have somehow managed to cope makes the next one more difficult, as increasing debt levels in many countries are approaching the limits of their creditworthiness. Because of the apparent success of recycling during the last three years, the awareness of this threat to the world's financial system has been dimmed; but a great deal of attention is now again being

given to it.

26. Let me refer to a recent warning by Arthur F. Burns [Chairman of the U.S. Federal Reserve Board], which he expressed before the Joint Economic Committee of Congress, where he stated that he had "communicated in strident tones" to leading bankers his concern about the risks of repayment of the $50 billion of loans that had been made by American banks to developing countries that are not oil producers.

27. Likewise, at a recent OECD meeting in Paris, the Chairman of the OECD Economic Policy Committee expressed a nagging worry about the accumulating current account deficits. He asserted that "assuming only moderate future oil price increases, the cumulative current account deficit of OECD countries between 1975-1980 could be estimated at $110 billion, and that of the non-oil poor countries at $160 billion."

28. Finally, Alan Greenspan, the former Chairman of President Ford's Council of Economic Advisors, stated last month before the Conference Board that at present world price relationships, the non-OPEC countries must, as a group, borrow at least $40 billion each year, cumulatively, year after year, with no way of shunting the cumulative debt on somebody else. . . . He concluded that "cumulating year after year, $40 billion deficits will eventually create such a huge debt structure that most Western industrial nations, and ultimately the U.S. itself, will find it difficult to meet the interest and amortization charges on the loans. . . . Realistically, either the real price of oil must come down or world oil use must fall dramatically."

29. There is no easy way out of this dilemma. It would be self-defeating to try to cover the balance of payment deficits by solutions that would lead to an over-expansion of world liquidity and would result in continuous and potentially rampant inflation—even though the debtor nations could then repay their existing obligations with a fraction of the real value of the amounts originally received. Nevertheless, the temptation for at least some countries to mitigate their debt

burden or to avoid default by passively accepting or even welcoming inflationary forces could prove to be dangerously attractive.

30. At the same time, however, oil prices would tend to rise with inflation; and the higher the rate of inflation, the greater the pressure on producing countries to restrict oil production rather than to accumulate financial reserves that might lose purchasing power through inflation. And with lower production, oil prices would increase even faster—thus triggering a downward slide of our economy at an ever increasing speed.

31. It is thus essential that the existing institutions such as the World Bank, the International Development Association, and the International Monetary Fund handle, where appropriate, the refinancing and new credit requirements of a large number of countries. This effort must be spearheaded and supplemented by agreements between the strong OECD and OPEC countries to supply the funds that are needed for the solution of the financial problems. What is required in many instances is a rescheduling of existing obligations, an abatement of interest rates, new soft loans and grants. In short, what we are really talking about are international arrangements for a sustained real transfer of wealth from the prosperous to the poorer countries, with its inevitable effect on the standard of living in the richer countries.

IV. What About the Future?

32. In looking now to our energy future, we face indeed most difficult and perhaps even intractable problems. On the one hand, there is the danger of declining energy supplies that would obstruct progress toward an expanding and prosperous world economy; on the other hand, increasing oil import costs might exceed the financial capability and the foreign exchange resouces of many oil-importing countries. But only broadly based economic progress would at least mitigate the balance of payment problems that oil imports would pose for many of them.

33. It would be foolish to deny that the future is clouded, that the task we have to undertake is very difficult, and the prospects for success at best uncertain. It is indeed not pleasant to arrive at so pessimistic a prognosis. But it would be wrong to interpret what we have said as an appraisal of the future that is immutable and inevitable. Yet only by frankly assessing the future problems that may confront us, and taking timely and decisive actions, do we provide ourselves with the opportunity to reverse the course of events and invalidate our earlier prognosis.

34. And with the immense stakes at issue, failure is not an acceptable option. It is thus incumbent upon us to marshal all available forces of human inventiveness and to accept the necessary economic sacrifices so that a shortage of energy will not extinguish the light at the end of the tunnel.

Notes

Chapter 1

1. U.S., Central Intelligence Agency, Office of Economic Research, *International Oil Development,* Washington, D.C., April 21, 1977. The *IOD* is an authoritative source of data and is used throughout this study along with other cited sources.

Chapter 2

1. For an important assessment of the demand-supply "balance" and attention to the critical financial implications of oil prices, see the appendix, Walter J. Levy, "U.S. Energy Policy in a World Context."

2. The company's forecast includes offshore to depths of 2,000 meters—farther out and deeper than is usually the case.

Chapter 3

1. Joseph A. Yager and Eleanor B. Steinberg, *Energy and U.S. Foreign Policy: A Report to the Energy Policy Project of the Ford Foundation* (Cambridge, Mass.: Ballinger Press, 1974). However, the entire quotation, including the "data," is profoundly misleading.

2. The size of coal deposits reflects the definitions given in the *Survey of Energy Resources:* total resources are the sum of known "reserves-in-place" and "additional" or inferred reserves; economically recoverable reserves are those exploitable by present techniques and conditions of price (these will always be less than "reserves-in-place" or additional, inferred reserves).

3. *Survey of Energy Resources,* p. 52.
4. Ibid., p. 28.
5. U.S., Federal Energy Administration, *National Energy Outlook—1976.*

Chapter 4

1. Generally, natural gas is piped to its destination; liquefied natural gas (LNG) implies the necessity to transport by tanker requiring compression of volumes, usually through reduction of temperature.
2. Data on natural gas reserves suffer from the same unreliability as oil's "authoritative" estimates. Forecasts of more than a few years ahead can be taken only to indicate trends and magnitudes.
3. British Petroleum, *LNG—the Next Ten Years.*

Chapter 10

1. For background and details of commitments accepted by member states, see Mason Willrick and Melvin A. Conant, "The International Energy Agency: An Interpretation and Assessment," *American Journal of International Law* 71, no. 2 (April 1977): 199-223.

Chapter 13

1. See Joseph P. Riva, Jr. and James E. Mielka, *Polar Energy Resources Potential,* Congressional Research Service (Washington, D.C.: Government Printing Office, September 1976).
2. Others view Soviet dependence upon imported oil as a likely and forbidding development—by the mid-eighties needing some 4-6 million barrels a day or even more. The Central Intelligence Agency report to the president of April 1977 contains this emphasis. It is our guess that the Soviets will not allow their energy autarky to lapse to such an extent without making extraordinary efforts to exploit their untapped resources—a crash effort that is not apparently under way.

Chapter 14

1. The authors are engaged in such a study under the auspices of the Seven Springs Center, an affiliate of Yale University.